Synthesis Lectures on Engineering, Science, and Technology

The focus of this series is general topics, and applications about, and for, engineers and scientists on a wide array of applications, methods and advances. Most titles cover subjects such as professional development, education, and study skills, as well as basic introductory undergraduate material and other topics appropriate for a broader and less technical audience.

Jerry D. Gibson

Information Theoretic Principles for Agent Learning

 Springer

Jerry D. Gibson
Department of Electrical and Computer
Engineering
University of California
Santa Barbara, CA, USA

ISSN 2690-0300 ISSN 2690-0327 (electronic)
Synthesis Lectures on Engineering, Science, and Technology
ISBN 978-3-031-65387-2 ISBN 978-3-031-65388-9 (eBook)
https://doi.org/10.1007/978-3-031-65388-9

This Springer imprint is published by the registered company Springer Nature Switzerland AG
The registered company address is: Gewerbestrasse 11, 6330 Cham, Switzerland

If disposing of this product, please recycle the paper.

To Elaine and Erin

Contents

Background and Overview

1

1.1 Introduction

A Learning Agent is a tool that is capable of learning from its experiences. With the explosion of research and applications for Artificial Intelligence, a need has developed for methods that provide insights into what could be happening inside the machine/deep learning black boxes, called explainability. A number of researchers have turned to information theory to explore the processes by which a learning agent acquires knowledge from large data sets and utilizes this knowledge in A.I. applications. More specifically, some researchers have backed away from trying to understand what is happening inside the machine/deep learning black boxes and considered what information is available to be learned from sequences with various properties.

So, the focus is on using a learning agent to analyze time series sequences of data [4, 6, 15, 16, 18, 19, 21, 22, 38, 38, 43]. Information theoretic principles play a large role in these analyses. However, practitioners in data science are not interested in the classical emphasis of Shannon information theory on fundamental limits of communication and compression and have not typically taken the standard information theory classes. Further, gleaning the needed information theory principles useful for analyses in data science are not easily mined from the standard communications and compression textbooks.

The purpose of this book is to provide the fundamentals of information theoretic techniques for statistical data science analyses and for characterizing the behavior and performance of a learning agent outside of the standard results on communications and compression fundamental limits.

J. D. Gibson, *Information Theoretic Principles for Agent Learning*, Synthesis Lectures on Engineering, Science, and Technology, https://doi.org/10.1007/978-3-031-65388-9_1

1.2   Classical Shannon Theory

Shannon developed information theory for the establishment of fundamental limits on communications, lossless compression, and lossy source compression. These fundamental limits involve the concepts of mutual information and entropy. Taken in isolation these probabilistic expressions have little to do with the physical world. However, by proving a coding theorem, Shannon showed that the maximum of the mutual information between the input and output of a communications channel described by a known set of conditional probabilities, the maximum taken over the set of possible input probabilities, has a physical interpretation as the maximum bit rate that can be transmitted over a channel with asymptotically small error probability. Shannon also demonstrated that the minimum bit rate needed to describe exactly, without any error, a discrete source with a known probability mass function is the discrete entropy. Then, he proved a lossy source coding theorem for a known source distribution that found the minimum rate to represent this source with an average distortion D is a minimum of the mutual information between the source and its reconstruction called the rate distortion function. In these cases entropy and mutual information became statistical quantities that represent performance for real world physical communication channels and sources. Most books on information theory and rate distortion theory explore these topics in great detail within the classical contexts of communications and compression [1–3, 25, 27].

1.3   Purpose of This Book

However, many researchers in data science and agent learning do not have the time nor inclination to study these books in detail and sort out what is useful to them outside of the applications to communications, lossless coding, and lossy source coding. The purpose of this book is to present information theoretic quantities, definitions, and results that provide or could provide insights into data science and learning, leaving interested researchers to access other details in the classical Shannon theory textbooks as desired.

1.4   Data Science, Agent Learning, and Information Theory

In data science, the goals are to analyze time series in order to identify structure and randomness, detect irregularities, identify patterns, discover change, and classify complexity. Entropy, cross entropy, mutual information, and Kullback-Leibler Divergence, also called relative entropy, are among the information theoretic tools that have been employed. Authors have also invoked classic results such as channel capacity and rate distortion theory, and their optimization problems, to uncover the behaviors of sequences. Some have even claimed analogies to topics not always in standard information theory textbooks such as successive refinement of information and source coding with side information.

Here, we develop fundamental information theoretic results and illustrate their applications in agent learning. We analyze how a learning agent, upon taking some observations of the environment, develops an understanding of the structure of the environment, formulates models of this structure, and studies any remaining apparent randomness or unpredictability. We see how unseen or unmodeled structure can be interpreted as unpredictability and even randomness.

1.5   Chapter Summaries

Chapter 2 introduces the concepts of entropy and mutual information for discrete random variables and develops the Kullback-Leibler Divergence, also called cross entropy or relative entropy, and defines the chain rules for these quantities. Information theoretic quantities for continuous random variables, namely differential entropy, entropy rate, entropy power, and maximum entropy, are presented in Chap. 3. These ideas are very useful in evaluating what can be learned from various time series.

The powerful idea of typical sequences is developed in Chap. 4, through the asymptotic equipartition property. This development includes both the weak law and strong laws of large numbers and jointly typical sequences. It is the author's view that these quantities are underutilized in the analyses of sequences. A cascade of systems is studied in Chap. 5 using Markov chains and entropy rate with particular examples for Gaussian and Laplacian random variables illustrating the strong connection between mean squared error estimation and entropy rate. These methods are employed later in Chaps. 7 and 8.

The important concept of a sufficient statistic is developed in Chap. 6 in the fields of hypothesis testing, estimation, and information theory. This is a unique collection of results not available as a group elsewhere. Chapter 7 introduces applications of entropy, entropy rate, and mutual information to agent learning and defines a specific quantity called Redundancy that has been used in such analyses. The analyses of cascaded systems in Chap. 8 extends the initial results in Chap. 5 by finding the minimum mean squared filtering, smoothing, and prediction errors in such systems and presents bounds on these errors using entropy power. Further applications of the prior-developed information theoretic ideas to the analyses of stationary Gaussian and autoregressive sequences are presented in Chap. 9, wherein a study of Code-Excited Linear Prediction (CELP) for speech coding is analyzed using mutual information and entropy power.

An idea that has been introduced and studied fairly extensively in the agent learning literature is the information bottleneck principle. This approach and the corresponding optimization problem are defined and discussed in Chap. 10. The similarity of the information bottleneck optimization to that of finding channel capacity and the rate distortion function, both classical and developed in later chapters is noted.

Information theoretic studies of agent learning have often invoked the terminology of channel capacity and rate distortion theory but outside of the classical fundamental limits on communications and compression. Chapter 11 contains an extensive treatment of channel capacity with developments of the basic ideas and results but without emphasizing the communications connection. This allows researchers in agent learning to have a firm foundation in these basic ideas so they can be correctly utilized in agent learning analyses. Chapter 12 is focused on rate distortion theory with treatments of the classical results but also introduces less studied ideas of the Shannon lower bound and the Shannon backward channel, successive refinement of information, and composite source models. These results and ideas have yet to find wide applications to agent learning but are likely to allow new insights and much more extensive results.

Entropy and Mutual Information

2.1 Introduction

Entropy, mutual information, and relative entropy are work horses in Shannon theoretic analyses for communications and compression. These quantities have already been utilized in the analyses of agent learning and the exploration of structure in sequences. This chapter provides concise treatments of these topics to establish notation and to summarize the basic results so that they are available for deeper study in later chapters. There is a beauty in these fundamental results that motivates further study and investigations.

2.2 Entropy

As an initial sequence model, we consider a discrete random variable U that takes on the values $\{1, 2, \ldots, M\}$, where the set of possible values of U is often called the *alphabet*, denoted as U, and the elements of the set are called *letters* of the alphabet. Let $P_U(u)$ denote the probability assignment over the alphabet, then we can define the *self-information* of the event $U = u$ by

$$I_U(u) = \log \frac{1}{P_U(u)} = -\log P_U(u) \ . \tag{2.1}$$

The quantity $I_U(u)$ is a measure of the information contained in the event $U = u$. Note that the base of the logarithm in Eq. (2.1) is unspecified. It is common to use base e, in which case $I_U(\cdot)$ is in natural units (nats), or base 2, in which case $I_U(\cdot)$ is in binary units (bits; that is, $H_b(X) = (\log_b a) H_a(X)$).

Either base is acceptable since the difference in the two bases is just a scaling operation. We primarily use base 2 in our discussions; therefore $I_U(\cdot)$ and related quantities will be in bits.

© The Author(s), under exclusive license to Springer Nature Switzerland AG 2025
J. D. Gibson, *Information Theoretic Principles for Agent Learning*, Synthesis Lectures on Engineering, Science, and Technology, https://doi.org/10.1007/978-3-031-65388-9_2

Let a random variable U have a sample space consisting of the set of all possible binary sequences of length N, denoted $\{1, 2, \ldots, 2^N\}$. If each of these sequences is equally probable, so that $P[u] = 2^{-N}$ for all u, the self-information of any event $U = u$ is

$$I_U(u) = -\log_2 P_U(u) = -\log_2\left(2^{-N}\right)$$
$$= N \text{ bits }.$$

The average or expected value of the self-information is called the *entropy,* also discrete entropy or absolute entropy, and is given by

$$H(U) = -\sum_{u=1}^{M} P_U(u) \log P_U(u) . \tag{2.2}$$

The following examples illustrate the calculation of entropy and how it is affected by probability assignments.

Given a random variable U with the alphabet $U = \{1, 2, 3, 4\}$ and probability assignments $P(1) = 0.8$, $P(2) = 0.1$, $P(3) = 0.05$, $P(4) = 0.05$, we calculate the entropy of U and compare the result to a random variable with equally likely values.

From Eq. (2.2),

$$H(U) = -\sum_{u=1}^{4} P_U(u) \log_2 P_U(u)$$
$$= -0.8 \log_2 0.8 - 0.1 \log_2 0.1 - 0.05 \log_2 0.05 -$$
$$0.05 \log_2 0.05$$
$$= 1.0219 \text{ bits }.$$

For equally likely values,

$$H(X) = -\log_2\left(\frac{1}{4}\right) = 2 \text{ bits }. \tag{2.3}$$

This last result illustrates the general property that for a discrete random variable U with possible values $\{u_1, u_2, \ldots, u_M\}$, then

$$H(U) \leq \log M \tag{2.4}$$

with equality if and only if the values of U are equally likely to occur.

Consider the random variables X and Y with joint pmf $p(x, y)$. Then

$$H(X, Y) = -\sum_{x \in X}\sum_{y \in Y} p(x, y) \log p(x, y) \tag{2.5}$$

$$= -E \log p(X, Y). \tag{2.6}$$

Further, with the conditional entropy $H(Y|X)$ defined as

$$H(Y|X) = \sum_{x \in X} p(x) H(Y|X = x) \tag{2.7}$$

$$= -\sum_{x \in X} p(x) \sum_{y \in Y} p(y|x) \log p(y|x) \tag{2.8}$$

$$= -\sum_{x}\sum_{y} p(x, y) \log p(y|x) \tag{2.9}$$

$$= -E_{p(x,y)} \log p(Y|X). \tag{2.10}$$

Expanding the joint entropy of (X, Y)

$$H(X, Y) = -\sum_{x}\sum_{y} p(x, y) \log p(x, y)$$

$$= -\sum_{x}\sum_{y} p(x, y) \left[\log p(y|x) + \log p(x)\right]$$

$$= H(X) + H(Y|X).$$

using the definitions. We can also show that $H(X, Y) = H(Y) + H(X|Y)$.

The definition of joint entropy and conditional entropy are natural since the joint entropy equals the entropy of one plus the conditional entropy of the other. We can continue the development as above to get the conditional version of the joint entropy

$$H(X, Y|Z) = H(X|Z) + H(Y|X, Z) . \tag{2.11}$$

Note that since $H(X, Y) = H(X) + H(Y|X) = H(Y) + H(X|Y)$, we have $H(X) - H(X|Y) = H(Y) - H(Y|X)$. This equality is useful in evaluating mutual information for various applications.

A property that is used often in communications system analysis is that conditioning cannot increase entropy. Later we show that this result follows from the expression for mutual information since $I(X; Y) = H(X) - H(X|Y) \geq 0$, so

$$H(X|Y) \leq H(X) \tag{2.12}$$

with equality if and only if X and Y are statistically independent.

The averaging is important for this property since individual terms may be greater, that is, $H(X|Y = y)$ may be greater than $H(X)$.

2.3   Mutual Information

We now consider two jointly distributed discrete random variables W and X with the probability assignment $P_{WX}(w, x)$, $w = 1, 2, \ldots, M$, $x = 1, 2, \ldots, N$. The *mutual information* over the joint ensemble is an important quantity defined by

$$
\begin{aligned}
I(W; X) &= \sum_{w=1}^{M} \sum_{x=1}^{N} P_{WX}(w, x) I_{W;X}(w; x) \\
&= \sum_{w=1}^{M} \sum_{x=1}^{N} P_X(x) P_{W|X}(w|x) P_W(w) \log \frac{P_{W|X}(w|x)}{P_W(w)}.
\end{aligned}
\tag{2.13}
$$

By a straightforward manipulation of the mutual information, we get

$$
I(W; X) = H(W) - H(W|X),
\tag{2.14}
$$

where $H(W|X)$ is the conditional entropy. Since entropy is a measure of uncertainty, we see from Eq. (2.14) that the mutual information can be interpreted as the average amount of uncertainty remaining about W after the observation of X.

As an example we calculate the mutual information for the probability assignments (with $M = 2$ and $N = 2$)

$$
P_W(1) = P_W(2) = \tfrac{1}{2}
\tag{2.15}
$$

and

$$
P_{X|W}(1|1) = P_{X|W}(2|2) = 1 - p
\tag{2.16}
$$

$$
P_{X|W}(1|2) = P_{X|W}(2|1) = p.
\tag{2.17}
$$

To calculate the mutual information, we use Eq. (2.13) so

$$
I(W; X) = I(X; W)
\tag{2.18}
$$

$$
= \sum_{w=1}^{2} \sum_{x=1}^{2} P_{WX}(w, x) \log \frac{P_{X|W}(x|w)}{P_X(x)}
\tag{2.19}
$$

$$
= 1 + (1 - p) \log(1 - p) + p \log p.
\tag{2.20}
$$

As another example, for the conditional probabilities $p(x|w)$ with $P_W(0) = \tfrac{1}{3}$ and $P_W(1) = \tfrac{2}{3}$, we find $H(W)$ and $H(W|X)$. The mutual information for these transition probabilities and the input probability assignment follows straightforwardly as

For $P_W(0) = 1/3$, $P_W(1) = 2/3$,

$$H(W) = -\frac{1}{3}\log\frac{1}{3} - \frac{2}{3}\log\frac{2}{3}$$

$$= 0.918 \text{ bits/letter.}$$

$$H(W|X) = -\sum_{x=1}^{3}\sum_{w=1}^{2} P_{W|X}(w|x) P_X(x) \log P_{W|X}(w|x) .$$

but

$$P_{W|X}(j|k) = \frac{P_{X|W}(k|j) P_W(j)}{P_X(k)} ,$$

so

$$H(W|X) = -\sum_{x=1}^{3}\sum_{w=1}^{2} P_{X|W}(x|w) P_W(w) \log P_{W|X}(w|x) .$$

We find that $P_X(0) = \frac{1}{3}$, $P_X(1) = 0.567$, $P_X(2) = 0.1$, and therefore, $H(W|X) = 0.5153$, so from Eq. (2.14)

$$I(W; X) \cong 0.918 - 0.515 = 0.403 \text{ bits/letter.}$$

A key property of mutual information is given by the following result, which is presented here without proof. Let W and X be jointly distributed random variables. The mutual information between W and X satisfies

$$I(W; X) \geq 0 \tag{2.21}$$

with equality if and only if W and X are statistically independent.

There are several other expressions for mutual information that can be useful in agent learning analyses. For example, the mutual information can be expanded in two ways involving entropy and conditional entropy,

$$I(X, Y) = H(X) - H(X|Y) \tag{2.22}$$

$$= H(Y) - H(Y|X) . \tag{2.23}$$

where the joint entropy of a pair of discrete rvs $(X, Y) \sim p(x, y)$ is defined as before.

2.4   Kullback-Leibler Divergence, Relative Entropy or Cross Entropy

The entropy of a random variable is a measure of the uncertainty; it is a measure of the information required on the average to describe the random variable. We have seen how entropy relates to mutual information, and now we consider the concept of relative entropy.

The *relative entropy*, *cross entropy* or Kullback Leibler distance [29] between two probability mass functions $p(x)$ and $q(x)$ is defined as

$$D(p \parallel q) = \sum_x p(x) \log \frac{p(x)}{q(x)} \tag{2.24}$$

$$= E_p \log \frac{p(x)}{q(x)} \tag{2.25}$$

In this definition, we use the convention (based on continuity arguments) that $O \log \frac{o}{q} = 0$ and $p \log \frac{p}{o} = \infty$.

We can show that

$$D(p|q) \geq 0 \tag{2.26}$$

with equality if and only if

$$p(x) = q(x) \quad \text{for all} \quad x . \tag{2.27}$$

This last result is called the Information Inequality and can be shown as follows.

We consider only the values of x such that $p(x) > 0$, called the support set of $p(x)$. Then

$$- D(p|q) = - \sum_{x \in A} p(x) \log \frac{p(x)}{q(x)} \tag{2.28}$$

$$\leq \log \sum_{x \in A} p(x) \log \frac{q(x)}{p(x)} \leq \log \sum_{x \in X} q(x) \tag{2.29}$$

$$= \log 1 = 0. \tag{2.30}$$

Since $\log t$ is a strictly concave function of t, we have equality iff $q(x)/p(x) = 1$ everywhere. Hence we have $D(p|q) = 0$ if and only if $p(x) = q(x)$ for all x.

We can use relative entropy to explore the properties of mutual information. For two random variables $(X, Y) \sim p(x, y)$ with marginals $p(x)$ and $p(y)$, the *mutual information* $I(X, Y)$ is the relative entropy between the joint distribution and the product distribution,

$$I(X, Y) = \sum_x \sum_y p(x, y) \log \frac{p(x, y)}{p(x)p(y)} \tag{2.31}$$

$$= D(p(x, y) \parallel p(x)p(y)) \tag{2.32}$$

$$= E_{p(x,y)} \log \frac{p(x, y)}{p(x)p(y)} . \tag{2.33}$$

The relative entropy leads to a direct proof of the non-negativity of mutual information. Explicitly, since $I(X; Y) = D(p(x, y) \parallel p(x)p(y)) \geq 0$ with equality if and only if $p(x, y) = p(x)p(y)$, i.e., X and Y are independent, then

$$I(X; Y) \geq 0 , \tag{2.34}$$

with equality if and only if X and Y are independent.

The relative entropy can be used to show that the uniform distribution over the range of X is the maximum entropy distribution over this range. Letting $u(x) = \frac{1}{|X|}$ be the uniform probability mass function (pmf) over X and $p(x)$ be the actual probability mass function for X, then we have directly that

$$D(p|u) = \sum p(x) \log \frac{p(x)}{u(x)} \tag{2.35}$$

$$= \log |X| - H(X), \tag{2.36}$$

but, we know that $D(p|u) \geq 0$, so

$$\log |X| \geq H(X).$$

This inequality is of fundamental importance.

2.5   Chain rules for Entropy and Mutual Information

We can build on the prior expressions and consider a set of jointly distributed random variables $X_1, X_2, \ldots, X_n$ drawn according to $p(x_1, x_2, \ldots, x_n)$. Their joint entropy is the sum of the conditional entropies as follows

$$H(X_1, X_2, \ldots, X_n) = \sum_{i=1}^{n} H(X_i | X_{i-1}, \ldots, X_1). \tag{2.37}$$

It is straightforward to justify this result since we already know that
$H(X_1, X_2) = H(X_1) + H(X_2|X_1)$,
so, [2]

$$H(X_1, X_2, X_3) = H(X_1, X_2) + H(X_3|X_1, X_2) \tag{2.38}$$

$$= H(X_1) + H(X_2|X_1) + H(X_3|X_1, X_2) \tag{2.39}$$

$$\vdots \quad = \quad \vdots \tag{2.40}$$

$$H(X_1, X_2, \ldots, X_n) = H(X_1) + H(X_2|X_1) \tag{2.41}$$

$$+ \ldots + H(X_n|X_{n-1}, \ldots, X_1) \tag{2.42}$$

$$= \sum_{i=1}^{n} H(X_i|X_{i-1}, \ldots, X_1) \tag{2.43}$$

where for $i = 1$, we define $H(X_1|X_0) = H(X)$.

The chain rule can also be obtained by starting with the basic definition of joint entropy and using conditional probabilities to expand the joint probability mass function. This is left as an exercise [2].

An important bound that finds wide applicability in practical applications is what is called the independence bound on entropy. For $X_1, X_2, \ldots, X_n$ with probability mass function $p(x_1, x_2, \ldots, x_n)$, we have by the chain rule for entropies,

$$H(X_1, X_2, \ldots, X_n) = \sum_{i=1}^{n} H(X_i | X_{i-1}, \ldots, X_1)$$

$$\leq \sum_{i=1}^{n} H(X_i)$$

since conditioning cannot increase entropy.

An important result that follows from the chain rule for entropy is that the mutual information also satisfies a chain rule. To see this we write

$$I(X_1, \ldots, X_n, Y) = H(X_1, \ldots, X_n) - H(X_1, \ldots, X_n | Y) \qquad (2.44)$$

$$= \sum_{i=1}^{n} H(X_i | X_{i-1}, \ldots, X_1) \qquad (2.45)$$

$$- \sum_{i=1}^{n} H(X_i | X_{i-1}, \ldots, X_1, Y) \qquad (2.46)$$

$$= \sum_{i=1}^{n} (H(X_i | X_{i-1}, \ldots, X_1) \qquad (2.47)$$

$$- H(X_i | X_{i-1}, \ldots, X_1, Y)) \qquad (2.48)$$

$$= \sum_{i=1}^{n} I(X_i, Y | X_{i-1}, \ldots, X_1), \qquad (2.49)$$

and thus we have

$$I(X_1, X_2, \ldots, X_n, Y) = \sum_{i=1}^{n} I(X_i, Y | X_{i-1}, \ldots, X_1) . \qquad (2.50)$$

We will see later that this can be used with an independence assumption or a Markov chain property to simplify further.

Differential Entropy, Entropy Rate, and Maximum Entropy

3

3.1 Introduction

The entropy associated with continuous random variables is distinct from the entropy of discrete random variables; in fact, it is sufficiently different in its properties that it is given a new name, *differential entropy*. Treatments of differential entropy are standard in basic information theory texts and in any first course in information theory. So, much of the material in Sect. 3.2 is widely available. The topics of entropy rate, entropy power, and maximum entropy, while usually contained in the same textbooks and introductory courses, are not as extensively emphasized as in this chapter and this book.

Further, the developments here are somewhat separated from information theoretic treatments of communications and compression and generalized so that the utility of differential entropy for exploring sequences as in agent learning is more evident.

3.2 Differential Entropy

Thus far we have defined the entropy and mutual information only for discrete random variables. Since continuous random variables play an important role in many applications, we develop these definitions and properties here.

Given an absolutely continuous random variable U with probability density function (pdf) $f_U(u)$ we define the *differential entropy* of U as

$$h(U) = -\int_{-\infty}^{\infty} f_U(u) \log f_U(u) \, du . \tag{3.1}$$

Although this expression appears quite similar to the expression for the entropy of a discrete random variable, there is a significant difference between the interpretations of absolute or discrete entropy and differential entropy. While $H(U)$ is an absolute indicator of "random-

© The Author(s), under exclusive license to Springer Nature Switzerland AG 2025 13
J. D. Gibson, *Information Theoretic Principles for Agent Learning*, Synthesis Lectures on Engineering, Science, and Technology, https://doi.org/10.1007/978-3-031-65388-9_3

ness," $h(U)$ is only an indicator of randomness with respect to a coordinate system; hence the names "absolute entropy" for $H(U)$ and "differential entropy" for $h(U)$. The following example illustrates the calculation of differential entropy and its property of indicating randomness with respect to a coordinate system.

Example

Consider an absolutely continuous random variable with uniform pdf

$$f_U(u) = \begin{cases} \dfrac{1}{a}, & \dfrac{-a}{2} \le u \le \dfrac{a}{2} \\ 0, & \text{elsewhere .} \end{cases} \tag{3.2}$$

(1) Letting $a = 1$ in Eq. (3.2) and we find $h(U)$. Then $h(U) = -\int_{-1/2}^{1/2} \log 1 \, du = 0$.
(2) For $a = 32$, we have $h(U) = 5$.
(3) Finally, with $a = \frac{1}{32}$, $h(U) = -5$.
The fact that differential entropy is a relative indicator of randomness is evident from these three special cases of the uniform distribution. Clearly, $h(U)$ is not an absolute indicator of randomness, since in case (3) $h(U)$ is negative, and negative randomness is difficult to interpret physically! The "reference" distribution is the uniform distribution over a unit interval, with "broader" distributions having a positive entropy and "narrower" distributions having a negative differential entropy.

    Example We calculate the differential entropy for an absolutely continuous random variable with pdf

$$f_U(u) = \begin{cases} \frac{1}{\alpha}e^{-u/\alpha}, & u > 0 \\ 0, & u \le 0, \end{cases}$$

From Eq. (3.1) with $\log = \ln$,

$$h(U) = -\int_0^{\infty} \frac{1}{\alpha} e^{-u/\alpha} \log \left[\frac{1}{\alpha} e^{-u/\alpha} \right] du$$

$$= \log \alpha + \left(\frac{1}{\alpha^2} \right) \alpha^2 e^{-u/\alpha} \left[-\frac{u}{\alpha} - 1 \right] \Big|_0^{\infty}$$

$$= \log e\alpha .$$

    The Laplacian density is often used as a model for transform coefficients for head and shoulders images and video in transform-based compression techniques that dominate still image and video coding today. We show that the differential entropy for a random variable U with the Laplacian pdf $f_U(u) = (1/\sqrt{2})e^{-\sqrt{2}|u|}$ for $-\infty < u < \infty$ is given by $h(U) = \log(e\sqrt{2})$. By direct substitution into Eq. (3.1), (using $\log_2$)

$$h(U) = -\int_{-\infty}^{\infty} \frac{1}{\sqrt{2}} e^{-\sqrt{2}|u|} \log \frac{1}{\sqrt{2}} e^{\sqrt{2}|u|} du$$

$$= -\log \frac{1}{\sqrt{2}} + \log_2 e = \log_2 e\sqrt{2} .$$

Given a Gaussian random variable U with mean μ_s and variance σ_s^2, it is straightforward to find an expression for its differential entropy $h(U)$. Using Eq. (3.1) and base 2 logs,

$$h(U) = -\int_{-\infty}^{\infty} f_U(u) \log f_U(u) du$$

$$= -\int_{-\infty}^{\infty} \frac{1}{\sqrt{2\pi}\sigma_s} e^{-(u-\mu_s)^2/2\sigma_s^2} \log \frac{1}{\sqrt{2\pi}\sigma_s} e^{-(u-\mu_s)^2/2\sigma_s^2} du$$

$$= \log \sigma_s \sqrt{2\pi e} .$$

The importance of the Gaussian distribution for finding appropriate performance bounds is emphasized by the following result.

For any absolutely continuous random variable ξ, the pdf that maximizes the differential entropy

$$h(\xi) = -\int f_\xi(\xi) \log f_\xi(\xi) \, d\xi \tag{3.3}$$

subject to the constraint that

$$\int_{-\infty}^{\infty} \xi^2 f_\xi(\xi) \, d\xi \leq \sigma_{\max}^2 \tag{3.4}$$

is

$$f_\xi(\xi) = \frac{1}{\sqrt{2\pi}\sigma_{\max}} e^{-\xi^2/2\sigma_{\max}^2} . \tag{3.5}$$

This result can be proved in several ways, including calculus of variations [7], relative entropy, and Jensen's inequality [27]; however, an alternative method is used here [1]. Let $f_\eta(\eta)$ be an arbitrary pdf that satisfies the constraint in Eq. (3.4), and let $f_\xi(\xi)$ be given by Eq. (3.5). Then

$$-\int_{-\infty}^{\infty} f_\eta(\alpha) \log f_\xi(\alpha) \, d\alpha \tag{3.6}$$

$$= \int_{-\infty}^{\infty} f_\eta(\alpha)(\log \sqrt{2\pi}\sigma_{\max} \tag{3.7}$$

$$+ \frac{\alpha^2}{2\sigma_{\max}^2} \log e) \, d\alpha \tag{3.8}$$

$$= \tfrac{1}{2} \log 2\pi e\sigma_{\max}^2. \tag{3.9}$$

Now consider

$$h(\eta) - \tfrac{1}{2} \log 2\pi e \sigma_{\max}^2 = \int_{-\infty}^{\infty} f_\eta(\alpha) \log \frac{f_\xi(\alpha)}{f_\eta(\alpha)} \, d\alpha$$

$$\leq \log e \int_{-\infty}^{\infty} f_\eta(\alpha) \left[\frac{f_\xi(\alpha)}{f_\eta(\alpha)} - 1 \right] d\alpha = 0 , \tag{3.10}$$

where the inequality follows from the fact that $\log \beta \leq (\beta - 1) \log e$. Thus,

$$h(\eta) \leq \tfrac{1}{2} \log 2\pi e \sigma_{\max}^2 \tag{3.11}$$

with equality if and only if $f_\xi(\alpha)/f_\eta(\alpha) = 1$ for all α. Hence, the desired result follows.

To reiterate, differential entropies are unlike absolute entropy in that differential entropy is not always positive and not necessarily finite. Another property is that differential entropy is not invariant to a one-to-one transformation of the random variable. We show this last fact in the following.

Given an absolutely continuous random variable X with pdf $f_X(x)$ and the transformation $Y = aX + b$, we find an expression for the entropy of Y in terms of $h(X)$. How has the transformation affected the result?

Given $f_X(x)$ and $Y = aX + b$, then we have,

$$f_Y(y) = \frac{f_X(x)}{|a|} \bigg|_{x = \frac{y-b}{a}} = \frac{f_X\left(\frac{y-b}{a}\right)}{|a|} . \tag{3.12}$$

Continuing from Eq. (3.1),

$$h(Y) = -\int_{-\infty}^{\infty} \frac{f_X\left(\frac{y-b}{a}\right)}{|a|} \log \frac{f_X\left(\frac{y-b}{a}\right)}{|a|} dy$$

$$= -\int_{-\infty}^{\infty} \frac{f_X\left(\frac{y-b}{a}\right)}{|a|} \left\{ \log f_X\left(\frac{y-b}{a}\right) - \log |a| \right\} dy .$$

Now,

$$\int_{-\infty}^{\infty} f_Y(y) \, dy = \int_{-\infty}^{\infty} \frac{f_X\left(\frac{y-b}{a}\right)}{|a|} \, dy = 1$$

and $x = \frac{y-b}{a}$, hence

$$-\int_{-\infty}^{\infty} \frac{f_X\left(\frac{y-b}{a}\right)}{|a|} \log f_X\left(\frac{y-b}{a}\right) dy = h(X) ,$$

so

$$h(Y) = h(X) + \log |a| .$$

The transformation has changed the differential entropy by an additive constant.

Consider the discrete random variable X with alphabet $\mathrm{X} = \{x_1, x_2, x_3, x_4\}$ each with their respective probability $P_X(x_1) = \frac{1}{2}$, $P_X(x_2) = \frac{1}{4}$, $P_X(x_3) = P_X(x_4) = \frac{1}{8}$, and the linear transformation $Y = aX + b$. We find $H(Y)$. What effect has the transformation had on the entropy of the discrete random variable X? The probability density function for Y is $P_Y(y_1) = P_Y(ax_1 + b) = \frac{1}{2}$, $P_Y(y_2) = P_Y(ax_2 + b) = \frac{1}{4}$, $P_Y(y_3) = P_Y(ax_3 + b) = \frac{1}{8}$, $P_Y(y_4) = P_Y(ax_4 + b) = \frac{1}{8}$. Thus, from Eq. (2.2),

$$H(Y) = -\sum_{j=1}^{4} P_Y(y_j) \log P_Y(y_j)$$

$$= -\sum_{j=1}^{4} P_Y(ax_j + b) \log P_Y(ax_j + b) .$$

But $P_Y(ax_j + b) = P_X(x_j)$, so

$$H(Y) = -\sum_{j=1}^{4} P_X(x_j) \log P_X(x_j) = H(X) .$$

The transformation has not changed the discrete entropy.

For a general multivariate probability density function of absolutely continuous random variables denoted by $f_X(x_1, x_2, \ldots, x_N)$, consider the one-to-one transformation represented by $y_i = g_i(X), i = 1, 2, \ldots, N$. We find an expression for the differential entropy of the joint pdf of Y, $f_Y(y_1, y_2, \ldots, y_N)$ in terms of the Jacobian of the transformation,

$$f_X(x_1, x_2, \ldots, x_N), \quad \text{and} \quad y_i = g_i(X), so$$

$$f_Y(y_1, y_2, \ldots, y_N) = \frac{f_X(x_1, x_2, \ldots, x_N)}{|J(x_1, x_2, \ldots, x_N)|}\bigg|_{x_j = x_{ji}} .$$

Assuming that there is only one solution to $y_i = g_i(X)$, then from Eq. (3.1),

$$h(Y) = E\{-\log f_Y(y_1, \ldots y_N)\}$$
$$= E\{-\log f_X(x_1, x_2, \ldots, x_N)\} + E\{\log |J|\}$$
$$= h(X) + E\{\log |J|\} ,$$

where $J = J(x_1, x_2, \ldots, x_N)$. Thus, the differential entropies differ by the expected value of the logarithm of the Jacobian.

The mutual information of two jointly distributed continuous random variables, say W and X, can also be defined as

$$I(W; X) = \int_{-\infty}^{\infty} \int_{-\infty}^{\infty} f_{WX}(w, x) \log \frac{f_{WX}(w, x)}{f_W(w) f_X(x)} \, dw \, dx$$
$$= I(X; W) . \tag{3.13}$$

As in the discrete case, the mutual information can be expressed in terms of differential (here) entropies as

$$I(W; X) = h(W) - h(W|X)$$
$$= h(X) - h(X|W) , \tag{3.14}$$

where

$$h(W|X) = -\int_{-\infty}^{\infty} \int_{-\infty}^{\infty} f_{WX}(w, x) \log f_{W|X}(w|x) \, dw \, dx . \tag{3.15}$$

Fortunately for our subsequent uses of $I(W; X)$, the mutual information is invariant under any one-to-one transformation of the variables, even though the individual differential entropies are not. We illustrate this by the following example.

Example We show that mutual information is invariant to one-to-one transformations. That is, given two continuous random vectors X and Z with mutual information $I(X; Z)$ and a one-to-one transformation $y_i = g_i(X), i = 1, 2, \ldots, N$, find $I(Y; Z)$. Straightforwardly, we know that

$$I(X; Z) = h(X) - h(X|Z)$$

and

$$I(Y; Z) = h(Y) - h(Y|Z) .$$

We are given that Y is a one-to-one transformation of X, so from a prior result

$$h(Y) = h(X) + E\{\log |J|\}$$

and

$$h(Y|Z) = h(X|Z) + E\{\log |J|\} .$$

Therefore,

$$I(Y; Z) = h(Y) - h(Y|Z)$$
$$= h(X) - h(X|Z) = I(X; Z) .$$

The mutual information is invariant under a one-to-one transformation since the term involving the Jacobian is in both differential entropy expressions and subtracts out.

In his landmark 1948 paper [7], Shannon defined the entropy power (also called entropy rate power) to be the power in a Gaussian white noise limited to the same band as the original ensemble and having the same entropy. He then used the entropy power in bounding the capacity of certain channels and for specifying a lower bound on the rate distortion function of a source. We develop the basic definitions and relationships concerning entropy power in the next section for use later in the book.

3.3   Entropy Power and Entropy Rate

Given a random variable X with probability density function $p(x)$, the differential entropy is

$$h(X) = -\int_{-\infty}^{\infty} p(x) \log p(x) dx \tag{3.16}$$

where X has the variance $var(X) = \sigma^2$. Since the Gaussian distribution has the maximum differential entropy of any distribution with mean zero and variance σ^2 [2],

$$h(X) \le \frac{1}{2} \log 2\pi e \sigma^2 \tag{3.17}$$

from which we obtain

$$Q = \frac{1}{(2\pi e)} \exp 2h(X) \le \sigma^2 \tag{3.18}$$

where Q was defined by Shannon to be the *entropy power* associated with the differential entropy of the original random variable X [7]. In addition to defining entropy power, this equation shows that the entropy power is the *minimum variance* that can be associated with the not-necessarily-Gaussian rv X and differential entropy $h(X)$.

Note that Eq. (3.18) allows us to calculate the entropy power associated with any given entropy. For example, if the random variable X is Laplacian [32] with parameter λ, then $h(X) = \ln(2\lambda e)$ and we can substitute this into Eq. (3.18) and solve for the entropy power $Q = 2e\lambda^2/\pi$. Since the variance of the Laplacian distribution is $\sigma^2 = 2\lambda^2$, we see that $Q < \sigma^2$, as expected from Eq. (3.18). This emphasizes the fact that the entropy power is not limited to Gaussian processes. This simple result is useful since speech signals as well as the linear prediction error for speech are often modeled by Laplacian distributions.

For an n-vector $\mathbf{X}$ with probability density $p(x^n)$, and covariance matrix $K_{\mathbf{X}} = E[(\mathbf{X} - E(\mathbf{X}))(\mathbf{X} - E(\mathbf{X}))^T]$, we have that

$$h(\mathbf{X}) \le \frac{1}{2} log[(2\pi e)^n |K_{\mathbf{X}}|] \tag{3.19}$$

from which we can construct the vector version of the entropy power as

$$Q_{\mathbf{X}} = \frac{1}{(2\pi e)^n} \exp 2h(\mathbf{X}) \le |K_{\mathbf{X}}|. \tag{3.20}$$

We can write a conditional version of Eq. (3.18) as

$$Q_{X|Y} = \frac{1}{(2\pi e)} \exp 2h(X|Y) \le Var(X|Y) \tag{3.21}$$

We will have the occasion to study pairs of random vectors $\mathbf{X}$ and $\mathbf{Y}$ where we use the vector $\mathbf{Y}$ to form the best estimate of $\mathbf{X}$. If $K_{\mathbf{X}|\mathbf{Y}}$ is the covariance matrix of the minimum mean squared error estimate of $\mathbf{X}$ given $\mathbf{Y}$, then we have

$$h(\mathbf{X}|\mathbf{Y}) \leq \frac{1}{2}log[(2\pi e)^n |K_{\mathbf{X}|\mathbf{Y}}|] \tag{3.22}$$

and from which we can get an expression for the conditional entropy power, $Q_{\mathbf{X}|\mathbf{Y}}$,

$$Q_{\mathbf{X}|\mathbf{Y}} = \frac{1}{(2\pi e)^n} \exp 2h(\mathbf{X}|\mathbf{Y}) \leq |K_{\mathbf{X}|\mathbf{Y}}|. \tag{3.23}$$

So, $Q_{\mathbf{X}|\mathbf{Y}}$ is upper bounded by the determinant of the conditional error covariance matrix, $|K_{\mathbf{X}|\mathbf{Y}}|$. We have equality in Eqs. (3.17)–(3.23) if the corresponding random variables or random vectors are Gaussian.

Often our interest is in investigating the properties of stationary random processes. Thus, if we let $\mathbf{X}$ be a stationary continuous-valued random process with samples $X^n = [X_i, i = 1, 2, ..., n]$, then the differential entropy rate of the process $\mathbf{X}$ is [31]

$$\bar{h} = \lim_{n\to\infty} \frac{1}{n} h(X^n) = \lim_{n\to\infty} h(X_n|X^{n-1}) \tag{3.24}$$

We assume that this limit exists in our developments and we drop the overbar notation and use $h = \bar{h}$. Using the entropy rate in the definition of entropy power yields the nomenclature *entropy rate power*.

Within the context of calculating and bounding the rate distortion function of a a discrete-time stationary random process, Kolmogorov [28] and Pinsker [30] derived an expression for the entropy rate power in terms of its power spectral density. If we now consider a discrete-time stationary Gaussian process with correlation function $\phi(k) = E[X_j X_{j+k}]$, the periodic discrete-time power spectral density is defined by

$$\Phi(\omega) = \sum_{-\infty}^{\infty} \phi(k) \exp{(j\omega k)} \tag{3.25}$$

for $|\omega| \leq \pi$. We know that an n-dimensional Gaussian density with correlation matrix Φ_n has the differential entropy $h(\mathbf{X}) = (n/2)\log{(2\pi e|\Phi_n|^{1/n})}$. Then, the entropy rate power Q can be found from [3, 26]

$$\log Q = \lim_{n\to\infty} \log |\Phi_n|^{1/n} \tag{3.26}$$

which yields [3, 26]

$$Q = \exp{[\frac{1}{2\pi} \int_{-\pi}^{\pi} \log \Phi(\omega) d\omega]} \tag{3.27}$$

as the entropy rate power.

Later we develop a closely related result for any distribution using the definition of entropy power.

3.4   Maximum Entropy

There are many problems where we may not know the distribution of the sequences we are studying but we do know other properties of the sequences, such as expected values or correlation properties. In this situation researchers often pursue the maximum entropy distribution since this approach assumes as little as possible about the sequences while still satisfying the known constraints.

To state the general maximum entropy problem, assume a continuous random variable X with $x \subseteq \Omega$. Then, the goal is to find the pdf $f_X(x)$ to maximize the differential entropy $h(X)$ subject to the conditions:

1. $f_X(x) \geq 0$ for $x \subseteq \Omega$ and $f_X(x) = 0$ outside of Ω,
2. $\int_\Omega f_X(x)dx = 1$
3. $\int_\Omega f_X(x)g_i(x)dx = r_i$ for $i = 1, 2, \cdots, K$

Therefore, the optimization problem involves forming the functional to be maximized, here defined as $J(f_X(x))$, as

$$J(f_X(x)) = - \int_\Omega f_X(x) \ln f_X(x) + \lambda_0 \int_\Omega f_X(x)dx \tag{3.28}$$

$$+ \sum_{i=1}^{K} \lambda_i \int_\Omega f_X(x)g_i(x)dx \tag{3.29}$$

Using regular calculus or the calculus of variations, it is straightforward to show that the necessary conditions for the maximum entropy distribution is that the pdf has the form

$$f_X(x) = e^{\lambda_0 - 1 + \sum_i \lambda_i g_i(x)} \tag{3.30}$$

where the λ_i are chosen to satisfy the constraints. Relative entropy can be used to show that this form uniquely maximizes $J(f_X(x))$ over all densities that satisfy the constraints [2].

We already know from Sect. 3.2 that the Gaussian distribution is the maximum entropy pdf for constraints on the mean and variance. It can be shown using the above results that the maximum entropy distribution for $\Omega = [0, \infty)$ and $EX = \mu$ is

$$f_X(x) = \frac{1}{\mu}e^{-\frac{x}{\mu}}, x \geq 0 \tag{3.31}$$

$$= 0, x < 0 \tag{3.32}$$

Typical Sequences and the AEP

4

4.1 Introduction

Agent learning is about observing sequences and trying to draw conclusions about structure that might be present and how this structure might be represented. The idea of typical sequences and a typical set are fundamentals in information theoretic developments of lossless source coding, channel capacity, and rate distortion theory. However, researchers in time series analysis and agent learning may not be familiar with these ideas.

In this chapter, typical sequences and the ideas of typical and nontypical sets are introduced and then extended to jointly typical sequences and sets. These ideas fall out of the basic results from the weak law of large numbers. It appears that, although typical sequences have not been exploited in agent learning analyses or to explore structure in sequences, there is much to be gained by translating these ideas into the agent learning problem.

4.2 Weak Law of Large Numbers

Let $X_1, X_2, \ldots,$ be a sequence of i.i.d. random variables with finite mean $EX = \mu$, then for $\varepsilon > 0 \; (M_n = \frac{1}{n} \sum_j X_j)$

$$\lim_{n \to \infty} P[|M_n - \mu| < \varepsilon] = 1. \tag{4.1}$$

The weak law states that for a large enough fixed values of n, the sample mean using n samples will be close to the true mean with high probability.

Chebychev's Inequality

Let Y be a random variable with mean μ and variance σ^2. Then

$$P_r\{|Y - \mu| > \epsilon\} \leq \frac{\sigma^2}{\epsilon^2}. \tag{4.2}$$

© The Author(s), under exclusive license to Springer Nature Switzerland AG 2025
J. D. Gibson, *Information Theoretic Principles for Agent Learning*, Synthesis Lectures on Engineering, Science, and Technology, https://doi.org/10.1007/978-3-031-65388-9_4

Weak Law of Large Numbers

Let $Z_1, Z_2, \ldots, Z_n$, be a sequence of i.i.d. random variables with mean μ and variance σ^2. Let

$$\bar{Z}_n = \frac{1}{n} \sum_{i=1}^{n} Z_i \ .$$

Then

$$P_r \left\{ \left| \bar{Z}_n - \mu \right| > \epsilon \right\} \leq \frac{\sigma^2}{n\epsilon^2} \ . \tag{4.3}$$

The weak law does not address what happens as a function of n. This question is taken up by the strong law.

4.3 Strong Law of Large Numbers

Let $X_1, X_2 \ldots$ be a sequence of i.i.d. rvs with $E[X] = \mu < \infty$ and finite variance, then

$$P \left[\lim_{n \to \infty} M_n = \mu \right] = 1 \ .$$

This says that w.p. 1 every sequence of sample mean calculations will eventually approach and stay close to $EX = \mu$.

4.4 Asymptotic Equipartition Property (AEP)

If $X_1, X_2, \ldots$, are i.i.d. $\sim p(x)$, then

$$\frac{-1}{n} \log p(X_1, \ldots, X_n) \longrightarrow H(X) \quad \text{in probability.} \tag{4.4}$$

Functions of independent random variables are also independent, so since the X_i are statistically independent, so are the $\log p(X_i)$. By Chebychev's inequality

$$P_r \left[\left| \frac{-1}{n} \log p(X_1, \ldots, X_n) - H(X) \right| \geq \varepsilon \right] \leq \tag{4.5}$$

$$\frac{\text{var} \left[\frac{-1}{n} \log p(X_1, \ldots, X_n) \right]}{\varepsilon^2}, \tag{4.6}$$

where

$$H(X) = -E \log p(X)$$

and

$$\text{var} \left[\frac{-1}{n} \log p(X_1, \ldots, X_n) \right] = \frac{n\sigma^2}{n^2} = \frac{\sigma^2}{n},$$

with

$$\sigma^2 = \sum_{x \in X} p(x)(\log p(x))^2 - \left(\sum_{x \in X} p(x) \log p(x)\right)^2$$

$$\therefore \ P_r\left[\left|\frac{-1}{n} \log p(X_1, \ldots, X_n) - H(X)\right| \geq \varepsilon\right] \leq \frac{\sigma^2}{n\varepsilon^2}$$

$$\therefore \ \lim_{n \to \infty} P_r[\quad] = 0 \,,$$

and the result follows.

By Chebychev's Inequality, with high probability,

$$\left|\frac{-1}{n} \log p(x_1, x_2, \ldots, x_n) - H(X)\right| \leq \varepsilon \,,$$

or

$$\frac{-1}{n} \log p(x_1, \ldots, x_n) - H(X) \leq \varepsilon$$

$$\frac{-1}{n} \log p(x_1, \ldots, x_n) \leq H(X) + \varepsilon$$

$$\log p(x_1, \ldots, x_n) \geq -n(H + \varepsilon)$$

or

$$p(x_1, \ldots, x_n) \geq 2^{-n(H+\varepsilon)} \tag{4.7}$$

or

$$\frac{1}{n} \log p(x_1, \ldots, x_n) + H(X) \leq \varepsilon$$

so

$$\frac{1}{n} \log p(x_1, \ldots, x_n) \leq -H(X) + \varepsilon$$

and

$$\log p(x_1, \ldots, x_n) \leq -n(H(X) - \varepsilon)$$

$$\therefore p(x_1, \ldots, x_n) \leq 2^{-n(H-\varepsilon)} \tag{4.8}$$

Note that (4.7) and (4.8) imply that the probability assigned to an observed sequence should be about 2^{-nH}.

This enables us to divide the set of all sequences into two sets, the *typical set*, where the sample entropy is close to the true entropy, and the *non-typical set*, which contains the other sequences. Typical sequences are where researchers in the analyses of deep learning should focus their attention. Any property that is proved for the typical sequence will then be true with high probability and will determine the average behavior of a large sample.

Let the random variable $X \in \{0, 1\}$ have a probability mass function defined by $p(1) = p$ and $p(0) = q$. If $X_1, X_2, \ldots, X_n$ are i.i.d. according to $p(x)$, then the probability of a sequence $x_1, x_2, \ldots, x_n$ is $\prod_{i=1}^{n} p(x_i)$. The probability of a particular sequence is $\sum X_i$. Then

$$p(X_1, X_2, \ldots, X_n) = p^{\sum X_i} q^{n - \sum X_i} ,$$

Thus, with $q = 1 - p, \sum X_i \cong np$, then,

$$p(X_1, X_2, \ldots, X_n) = p^{np}(1 - p)^{n(1-p)} ,$$

so

$$\log p(X_1, X_2, \ldots, X_n) = np \log p + n(1 - p) \log(1 - p)$$
$$= -nH(X),$$

we are simply saying that the number of 1's in the sequence is close to np (with high probability), and all such sequences have (roughly) the same probability $2^{-nH(p)}$.

From Blahut [25], "Whereas weakly typical sequences display approximately the right apparent entropy, strongly typical sequences display approximately the right relative frequency of symbols."

Definition of the Typical Set [2] The typical set $A_\epsilon^{(n)}$ with respect to $p(x)$ is the set of sequences $(x_1, x_2, \ldots, x_n) \in X^n$ with the following property

$$2^{-n(H(X)+\epsilon)} \leq p(x_1, x_2, \ldots, x_n) \leq 2^{-n(H(X)-\epsilon)} \tag{4.9}$$

As a consequence of the AEP, we can show that the set $A_\epsilon^{(n)}$ has the following properties:
1. If $(x_1, x_2, \ldots, x_n) \in A_\epsilon^{(n)}$,
then $H(X) - \epsilon \leq \frac{-1}{n} \log p(x_1, x_2, \ldots, x_n) \leq H(X) + \epsilon$.
2. $Pr\left\{A_\epsilon^{(n)}\right\} > 1 - \epsilon$ for n sufficiently large.
3. $\left|A_\epsilon^{(n)}\right| \leq 2^{n(H(X)+\epsilon)}$, where $|A|$ denotes the number of elements in the set A.
4. $\left|A_\epsilon^{(n)}\right| \geq (1 - \epsilon)2^{n(H(X)-\epsilon)}$, for n sufficiently large.
Thus, the typical set has probability nearly 1, all elements of the typical set are nearly equiprobable, and the number of elements in the typical set is nearly 2^{nH}.

4.5 Jointly Typical Sequences

One particularly useful tool in analyzing learning in sequences is the *jointly typical set.* The set $A_\epsilon^{(n)}$ of jointly typical sequences $\{(x^n, y^n)\}$ with respect to the distribution $p(x, y)$ is the set of n-sequences with empirical entropies within ϵ of the true entropies, i.e., [2]

$$A_\epsilon^{(n)} = \left\{ (x^n, y^n) \in X^n \times Y^n : \right. \tag{4.10}$$

$$\left| \frac{-1}{n} \log p(x^n) - H(X) \right| < \epsilon , \tag{4.11}$$

$$\left| \frac{-1}{n} \log p(y^n) - H(Y) \right| < \epsilon , \tag{4.12}$$

$$\left. \left| \frac{-1}{n} \log p(x^n, y^n) - H(X, Y) \right| < \epsilon \right\} , \tag{4.13}$$

where

$$p(x^n, y^n) = \prod_{i=1}^{n} p(x_i; y_i) . \tag{4.14}$$

Given this definition, the following results can be developed similarly to before.

Joint AEP [2] Let (X^n, Y^n) be sequences of length n drawn i.i.d. according to $p(x^n, y^n) = \prod_{i=1}^{n} p(x_i, y_i)$. Then

1. $\Pr((X^n, Y^n) \in A_\epsilon^{(n)}) \to 1$ as $n \to \infty$.
2. $|A_\epsilon^{(n)}| \leq 2^{n(H(X,Y)+\epsilon)}$.
3. If $(\tilde{X}^n, \tilde{Y}^n) \sim p(x^n)p(y^n)$, i.e., $\tilde{X}^n$, and $\tilde{Y}^n$ are independent with the same marginals as $p(x^n, y^n)$, then

$$\Pr((\tilde{X}^n, \tilde{Y}^n) \in A_\epsilon^{(n)}) \leq 2^{-n(I(X;Y)-3\epsilon)} . \tag{4.15}$$

Also, for sufficiently large n,

$$\Pr((\tilde{X}^n, \tilde{Y}^n) \in A_\epsilon^{(n)}) \geq (1 - \epsilon) 2^{-n(I(X;Y)+3\epsilon)} .$$

The set of jointly typical sequences is often used in proving Shannon's channel coding theorem. However, the jointly typical set can also provide insight into the relationship between two sequences in learning applications.

4.6   Perplexity

Entropy plays a role in language models for speech recognition. Consider a vocabulary of size $|V|$. For an application where the word sequence is completely random and thus the words are uniformly distributed, then the entropy is maximum at $\log_2 |V|$ bits. However, the numbers of words that can follow a particular word sequence is constrained by the language model.

More specifically, a language model can be described by a set of conditional probabilities $P(w_k|w_{k-1}, w_{k-2}, \cdots, w_{k-N+1})$ which is an N gram model of the grammar. Then, the probability of a specific word sequence is $P(W) = P(w_1, w_2, \ldots, w_K)$ is given by

$$P(w_1, w_2, \ldots, w_K) = \prod_{k=1}^{K} P(w_k|w_{k-1}, w_{k-2}, \ldots, w_{k-N+1}) \tag{4.16}$$

For this length K sequence, the entropy per word is

$$H_K = -\frac{1}{K} \log_2 P(w_1, w_2, \ldots, w_K). \tag{4.17}$$

The complexity of a language model is often characterized as the mean number of transitions leaving a node or the average branching factor of the language model and that is defined as the *Perplexity* given by

$$B_K = 2^{H_K} \tag{4.18}$$

for the length K sequence of word. If the sequence is ergodic, then we can consider

$$B = \lim_{K \to \infty} B_K = 2^H \tag{4.19}$$

B_K and B are often used in quantifying the recognition complexity of a language model.

For more details on perplexity and language models see the references [45, 46].

Markov Chains and Cascaded Systems

5

5.1 Introduction

Markov chains are indispensable models for cascaded systems in signal processing, communications, and information theory. Information theoretic descriptions of systems using Markov chains play a dominant role in communications and compression applications. In this chapter, we provide a brief introduction to results from communications and compression but in a more general context. Plus, we relate the classical measure of performance in estimation and prediction, mean squared error, to information theoretic descriptions of Markov chain models of cascaded systems.

5.2 Cascaded Signal Processing

Figure 5.1 shows a cascade of N signal processing operations with the estimator blocks at the output of each stage as studied by Messerschmitt [33]. He used the conditional mean at each stage and the corresponding conditional mean squared errors to obtain a representation of the distortion contributed by each stage. We analyze the cascade connection in terms of information theoretic quantities, such as mutual information, differential entropy, and entropy rate power. Similar to Messerschmitt, we consider systems that have no hidden connections between stages other than those explicitly shown. Therefore, we conclude directly from the data processing inequality [2] that

$$I(X; Y_1) \geq \ldots \geq I(X; Y_{N-1}) \geq I(X; Y_N) \geq I(X; \widehat{X}). \tag{5.1}$$

Since $I(X; Y_n) = h(X) - h(X|Y_n)$, it follows from Eq. (5.1) that for non-negative $h(\cdot)$,

$$h(X|Y_1) \leq \ldots \leq h(X|Y_{N-1}) \leq h(X|Y_N) \leq h(X). \tag{5.2}$$

© The Author(s), under exclusive license to Springer Nature Switzerland AG 2025 29
J. D. Gibson, *Information Theoretic Principles for Agent Learning*, Synthesis Lectures on Engineering, Science, and Technology, https://doi.org/10.1007/978-3-031-65388-9_5

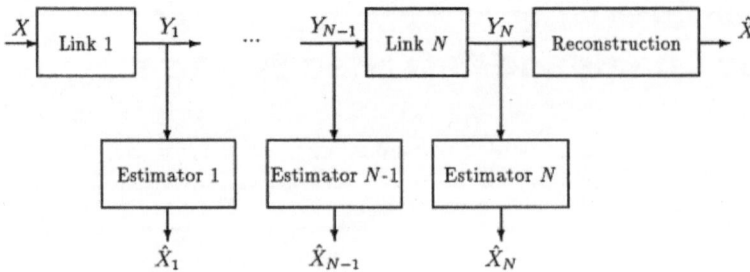

Fig. 5.1 N-link system block diagram (adapted from [33])

For the optimal estimators at each stage, the basic data processing inequality also yields $I(X; Y_n) \geq I(X; \widehat{X}_n)$, and thus $h(X|Y_n) \leq h(X|\widehat{X}_n)$.

These are the fundamental results, that additional processing cannot increase the mutual information.

Now we notice that the series of inequalities in Eq. (5.2) along with the entropy power expression in Eq. (3.18) gives us the series of inequalities in terms of entropy power at each stage in the cascaded signal processing operations:

$$Q_{X|Y_1} \leq \cdots \leq Q_{X|Y_{N-1}} \leq Q_{X|Y_N} \leq Q_X. \tag{5.3}$$

We can also write that

$$Q_{X|Y_n} \leq Q_{X|\widehat{X}_n}. \tag{5.4}$$

In the context of Eq. (5.3), the notation $Q_{X|Y_n}$ denotes the minimum variance when reconstructing an approximation to X given the sequence at the output of stage n in the chain [36].

5.3   Log Ratio of Entropy Powers

We can use the definition of the entropy power in Eq. (3.18) to express the logarithm of the ratio of two entropy powers in terms of their respective differential entropies as

$$\frac{1}{2} \log \frac{Q_X}{Q_Y} = [h(X) - h(Y)] \tag{5.5}$$

We are interested in studying the change in the differential entropy brought on by different signal processing operations by investigating the log ratio of entropy powers. However, in order to calculate the entropy power, we need an expression for the differential entropy! So, why do we need the entropy power?

First, entropy power may be easy to calculate in some instances, as we show later. Second, the accurate computation of the differential entropy can be quite difficult and requires considerable care [47]. Generally, the approach is to estimate the probability density function (pdf) and then use the resulting estimate of the pdf in Eq. (3.16) and numerically evaluate the integral.

Depending on the method used to estimate the probability density, the operation requires selecting bin widths, a window, or a suitable kernel [47], all of which must be done iteratively to determine when the estimate is sufficiently accurate. The mutual information is another quantity of interest, as we shall see, and the estimate of mutual information also requires multiple steps and approximations [40, 43]. These statements are particularly true when the signals are not i.i.d. and have unknown correlation.

In the following we consider special cases where Eq. (5.5) holds with equality when the entropy powers are replaced by the corresponding variances. We then provide justification for using the signal variance rather than entropy power in other situations. The Gaussian and Laplacian distributions often appear in studies of speech processing and other signal processing applications, so we show that substituting the variances for entropy powers for these distributions satisfies Eq. (5.5) exactly.

5.3.1   Gaussian Distributions

For two i.i.d. Gaussian distributions with zero mean and variances σ_X^2 and σ_Y^2, we have directly that $Q_X = \sigma_X^2$ and $Q_Y = \sigma_Y^2$, so

$$\frac{1}{2} \log \frac{Q_X}{Q_Y} = \frac{1}{2} \log \frac{\sigma_X^2}{\sigma_Y^2} = [h(X) - h(Y)] \tag{5.6}$$

which satisfies Eq. (5.5) exactly. Of course, since the Gaussian distribution is the basis for the definition of entropy power, this result is not surprising.

5.3.2   Laplacian Distributions

For two i.i.d. Laplacian distributions with parameters λ_X and λ_Y [32], their corresponding entropy powers $Q_X = 2e\lambda_X^2/\pi$ and $Q_Y = 2e\lambda_Y^2/\pi$, respectively, so we form

$$\begin{aligned}
\frac{1}{2} \log \frac{Q_X}{Q_Y} &= \frac{1}{2} \log \frac{\lambda_X^2}{\lambda_Y^2} \\
&= [\ln(2e\lambda_X) - \ln(2e\lambda_Y)] \\
&= [h(X) - h(Y)].
\end{aligned} \tag{5.7}$$

so the Laplacian distribution also satisfies Eq. (5.5) exactly. We thus conclude that we can substitute the variance, or for zero mean Laplacian distributions, the mean squared value for the entropy power in Eq. (5.5) and the result is the difference in differential entropies.

Interestingly, it can be shown that the Logistic distribution [32] also satisfies Eq. (5.5) exactly. In fact, random variables whose differential entropy has the form

$$h(X) = \frac{1}{2}log[A\sigma_X^2]$$

(5.8)

satisfy Eq. (5.5) exactly. Therefore, in the log ratio of entropy powers for cascaded systems, we do not have to explicitly calculate the entropy power since we can use the variance or mean squared error for these distributions in these expressions to find the mutual information gain or loss.

Hypothesis Testing, Estimation, Information, and Sufficient Statistics

6

6.1 Introduction

To analyze data in applications, it is necessary to form statistics that allow us to draw conclusions about what the data represents. Machine/deep learning approaches also form statistics within their black boxes and in order to understand these methods, we need to consider what statistics those learning algorithms might be using. In both cases, we need to determine whether the statistic being used is sufficient to draw conclusions.

Definitions of Sufficient Statistics take a slightly different form depending upon whether the problem is one of hypothesis testing, estimation, or information theory. We explore each of these definitions in this chapter and develop the basic results.

6.2 Definition of a Sufficient Statistic

To state a definition of a sufficient statistic, consider a probability mass function (pmf) that is indexed by a parameter θ, so that we have a family of pmfs $f_\theta(x)$. We take a sample X from one distribution in the family and form the statistic $S(X)$ from that sample. We hope to draw conclusions about θ from the statistics of the sample, here denoted by $S(X)$. The question is whether we have chosen a good statistic to draw conclusions about θ; that is, does $S(X)$ have all of the information that we need to estimate θ? This can be checked by determining if $S(X)$ removes any additional dependence of the distribution of X on the parameter θ. If so, then $S(X)$ is a sufficient statistic for the parameter θ.

While the stated definition is clear enough verbally, we need to develop a more specific definition for what is a sufficient statistic and study how the definition impacts applications in hypothesis testing, estimation, and information theory.

© The Author(s), under exclusive license to Springer Nature Switzerland AG 2025
J. D. Gibson, *Information Theoretic Principles for Agent Learning*, Synthesis Lectures on Engineering, Science, and Technology, https://doi.org/10.1007/978-3-031-65388-9_6

6.3   The Factorization Theorem and Implications

To restate the problem and set notation, we consider a family of pmfs $f_\theta(x)$ indexed by a parameter θ, and this family characterizes the possible distributions of a random variable X. We observe a value x of the random variable X, the value x is the data upon which we will draw conclusions about θ. Denote the statistic based on the data as $S(X).S(X = x)$ is what we use to estimate the parameter θ.

The Factorization Theorem [14] A statistic S is sufficient for θ if and only if there are functions g_θ and h such that

$$f_\theta(x) = g_\theta[S(x)]h(x) \tag{6.1}$$

The result is that rather than using the raw sample values x directly, we have formed rather some statistic $S(x)$ that is some aggregation or transformation of the sample values. If the $S(x)$ is a sufficient statistic, then this aggregation reduces the data to be processed but without losing any of the needed information to draw conclusions about θ.

Interestingly, this idea can be captured by the following scenario. Once we have the value of the statistic $S(x)$, it is possible to regenerate the original distribution $f_\theta(x)$ by forming a random quantity, say $\widehat{X}$, by randomly exciting the conditional distribution $f(X|S = s)$, so that $F_\theta(\widehat{X}) = F_\theta(X)$, no matter what value of unknown θ generated the original data samples.

It can be stated broadly that a sufficient statistic reduces the amount of data to be processed without losing any essential information. For a given problem, there can be many sufficient statistics which contain the same information but each perhaps providing a different amount of data reduction.

There is also the concept of a minimal sufficient statistic, which provides the greatest reduction of the data of all sufficient statistics. To be more specific, if S is a sufficient statistic and $S = G(T)$, then T is also a sufficient statistic. Knowing T thus allows us to know S through the mapping $G(\cdot)$, and if $G(\cdot)$ is one-to-one, S and T are said to be equivalent in that they provide the same amount of information. In these terms, if T is any sufficient statistic and there exists a function $G(\cdot)$ such that $S = G(T)$, then S is a minimal sufficient statistic.

6.4   Hypothesis Testing

In binary hypothesis testing, we are interested in testing an hypothesis H corresponding to the pdf $f(x|H)$ versus the alternative hypothesis K with corresponding pdf $f(x|K)$. The fundamental statistic that appears in this problem is the likelihood ratio

$$L(x) = \frac{f(x|K)}{f(x|H)} \tag{6.2}$$

and this likelihood ratio is treated as a sufficient statistic. In our current development, we can justify this by checking to see if the likelihood ratio in Eq. (6.2) satisfies the Factorization Theorem.

To illustrate this, note that we can write $f(x|K)$ as

$$\frac{f(x|K)}{f(x|H)} f(x|H) = L(x) f(x|H) \tag{6.3}$$

Setting $h(x) = f(x|H)$ and $S(x) = f(x|K)/f(x|H)$ and defining $g_\theta(s) = 1$, if H is true, and $g_\theta(s) = L(x)$, if K is true, then the likelihood ratio satisfies the Factorization Theorem [14] and $L(x)$ is a sufficient statistic for the binary hypothesis testing problem.

6.5   Estimation

As before, we consider a set of pmfs $f_\theta(x)$ indexed by θ and we wish to develop an estimator of some function of θ, say $g(\theta)$, based on observations of the random variable X. Thus, we take a sample x of the random variable X and form an estimate $Y(x)$ which we hope is close to the function $g(\theta)$. Note that thus far, we have not specified a loss function, in particular, since in current day machine/deep learning problems it is not desirable to choose a loss function.

We do need a measure of the performance of the estimator, however, so if $Y(x)$ takes on the value y, we specify a loss function $\phi(\theta, y)$ averaged over all values of θ to obtain [13]

$$\Phi(\theta, Y) = E_\theta[\phi(\theta, Y)] \tag{6.4}$$

The goal of our estimation problem then is to find an estimator $Y(\cdot)$ that minimizes this average loss function for all values of θ.

Of course, given the observations $X = x$, it is desirable that $Y(x)$ forms a sufficient statistic for estimating $g(\theta)$. Furthermore, our preference is for $Y(X)$ to be a minimal sufficient statistic for estimating $g(\theta)$, where both a sufficient statistic and a minimal sufficient statistic is as developed in Sect. 6.3.

Interestingly, if the estimator is the result of imposing a convex loss function to obtain an estimate, say $\eta(S)$, where S is a sufficient statistic for $f_\theta(x)$, then

$$\Phi(\theta, \eta) < \Phi(\theta, Y). \tag{6.5}$$

6.6   Information Theory

Again, we consider a family of pmfs $f_\theta(x)$ indexed by a parameter θ, and this family characterizes the possible distributions of a random variable X. We observe a value x of the random variable X, the value x is the data upon which we will draw conclusions about θ.

Denote the statistic based on the data as $S(X)$. We want to develop information theoretic conditions for $S(X)$ to be a sufficient statistic for the parameter θ or some function of θ, $g(\theta)$ [2].

We can characterize the processing in terms of the Markov Chain

$$\theta \to X \to S(X) \tag{6.6}$$

so by the Data Processing Inequality

$$I(\theta; S(X)) \leq I(\theta; X) \tag{6.7}$$

and no information is lost with equality. $S(X)$ is said to be sufficient for θ when equality holds in Eq. (6.7).

The following definition gives a specific condition for $S(X)$ to be called a sufficient statistic. **Definition**: A function $S(X)$ is said to be a **sufficient statistic** for θ if X is independent of θ given $S(X)$ or if the following Markov Chain holds

$$\theta \to S(X) \to X \tag{6.8}$$

If Eq. (6.6) holds then $I(\theta; X|S(X)) = 0$.

We can show this last result by starting with the original Markov Chain in Eq. (6.6) and expanding $I(\theta; X, S(X))$ in two ways

$$\begin{aligned} I(\theta; X, S(X)) &= I(\theta; X) + I(\theta; S(X)|X) \\ &= I(\theta; S(X)) + I(\theta; X|S(X)) \end{aligned} \tag{6.9}$$

The quantity $I(\theta; S(X)|X)$ can be expanded as

$$I(\theta; S(X)|X) = H(\theta|X) - H(\theta|X, S(X)) = 0, \tag{6.10}$$

since $H(\theta|X, S(X)) = H(\theta|X)$ from the Markov Chain in Eq. (6.6).

It is also evident that $I(\theta; X|S(X)) \geq 0$, so combining this with Eq. (6.10) in Eq. (6.9), we can conclude that

$$I(\theta; X) \geq I(\theta; S(X)) \tag{6.11}$$

with equality iff $I(\theta; X|S(X)) = 0$ or when

$$\theta \to S(X) \to X \tag{6.12}$$

forms a Markov Chain, which establishes the stated definition.

Therefore, $S(X)$ is a sufficient statistic for θ.

Information Theoretic Quantities and Learning

7

7.1 Introduction

The focus in this chapter is on using information theoretic ideas and results to analyze agent learning wherein upon taking some observations of the environment, we develop an understanding of the structure of the environment, formulate models of this structure, and study any remaining apparent randomness or unpredictability.

Results from agent learning have made use of the information theoretic ideas in Chaps. 2 and 3, and have created variations on those information theoretic ideas to capture particular characteristics that are distinct to agent learning problems. We summarize a few of these variations and newly defined quantities here.

7.2 Entropy and Entropy Rate in Agent Learning

In the agent learning literature, it is desired to explore the broad ideas of unpredictability and apparent randomness [4, 6]. For example, given a continuous valued sequence X_i, $i = 1, 2, \ldots, N$, for $x \in X$, how do we characterize whether this sequence is random or it has some structure?

To address such questions, it is common to investigate the *total Shannon entropy* of length-N sequences given by

$$h(X^{(N)}) = -\int P_X^{(N)}(X) \log P_X^{(N)}(X) dX^N \tag{7.1}$$

as a function of N to characterize learning. The name total Shannon entropy is appropriate since it is not the usual per component entropy of interest in lossless source coding [2], for example.

© The Author(s), under exclusive license to Springer Nature Switzerland AG 2025
J. D. Gibson, *Information Theoretic Principles for Agent Learning*, Synthesis Lectures on Engineering, Science, and Technology, https://doi.org/10.1007/978-3-031-65388-9_7

In association with the idea of learning or discerning structure in an environment, the *entropy gain* is defined as the difference between the entropies of length N and length $N - 1$ sequences as [4]

$$\Delta H(N) = h(X^N) - h(X^{N-1}) \tag{7.2}$$

Equation (7.2) was derived and studied much earlier by Shannon [7] not as an entropy gain but as a conditional entropy.

In particular, Shannon [7] defined the conditional entropy of the next symbol when the $N - 1$ preceding symbols are known as

$$h(X_N|X^{N-1}) = h(X_N, X^{N-1}) - h(X^{N-1}) = h(X^N) - h(X^{N-1}) \tag{7.3}$$

which is exactly Eq. (7.2); so the entropy gain from the agent learning literature is simply the conditional entropy expression developed by Shannon in 1948.

Elias [37] considered the conditional entropy introduced by Shannon and called it the *entropy added by the Nth term*, which again is consistent with the designation of *entropy gain* in the agent learning literature as in Eq. (7.2). Elias desired to find an upper bound on this added entropy. Noting that the differential entropy of an Nth order Gaussian sequence is given by $\frac{1}{2} \log \left[2\pi e |\mathcal{R}_N|^{1/N} \right]$, Elias shows that the entropy added by the Nth term is

$$\Delta H(N) = \frac{1}{2} \log 2\pi e \frac{|\mathcal{R}_{N+1}|}{|\mathcal{R}_N|} \tag{7.4}$$

Going beyond the concept of entropy gain, a definition of *information gain*, represented by $\Delta H(N)$ and expressed as a relative entropy has been offered and studied by Crutchfield and Feldman [4, 6] as

$$\Delta H(N) = D(P_X^{(N)}(X) || P_X^{(N-1)}(X)) \tag{7.5}$$

In Eq. (7.5), the support set of the two distributions is not the same, so the $P_X^{(N-1)}(X)$ is extended by concatenating all values of the x_N symbol with the prior symbols $x_0, x_1, \ldots, x_{N-1}$ with equal probability [4].

It is also shown in [4] that (this result is in Shannon [7] and Elias [37] as well)

$$\overline{h} = \lim_{N \to \infty} \Delta H(N) \tag{7.6}$$

which is the definition of differential entropy rate stated in Chap. 3, and where we let $\overline{h} = h(\mathcal{X})$ for notational compactness and to be consistent with [4].

7.3   Redundancy in Agent Learning

Further, in the learning literature, two definitions of a quantity called *Redundancy* are offered. One definition is as the difference between the maximum value of the entropy rate $\log |\mathcal{X}|$, where $|\mathcal{X}|$ is the cardinality of a discrete alphabet or the volume of the support set for a continuous variable, and the entropy rate $\overline{h}$ so that the redundancy is [4]

$$R = \log |\mathcal{X}| - \overline{h} \tag{7.7}$$

A second definition of redundancy $D_N(P_X^{(N)}(X)||U^{(N)})$ is the relative entropy between the known distribution $P_X^{(N)}$ and the uniform distribution, $U^{(N)}$, asymptotically in N,

$$R = \lim_{N \to \infty} D_N(P_X^{(N)}(X)||U^{(N)}) \tag{7.8}$$

Thus, for the definitions of redundancy in Eqs. (7.7) and (7.8), it can be stated that the redundancy R is an indicator of the information gained when an agent learns that the actual distribution is different from an uniform distribution as the sequence length becomes asymptotically large.

To study how the redundancy evolves with finite length N observations of the environment, a version of the redundancy, called N-*redundancy*, is defined if the actual distribution of the length N sequence is known to be $P_X^{(N)}$, so the entropy is $h(X_1, \ldots, X_N)$ and [4]

$$R(N) \equiv h(X_1, \ldots, X_N) - N\overline{h} \tag{7.9}$$

Equations (7.8) and (7.9) are special cases of the generalized definition of redundancy from information theory which is the difference between the expected length of a lossless code and the lower limit for the expected length of the code, expressed in terms of a relative entropy [2].

A characterization of the per symbol entropy when N observations of the environment are available compared to the per symbol entropy with an infinite number of measurements is given by the *per symbol N-redundancy* defined as

$$r(N) = \Delta R(N) = \Delta H(N) - \overline{h} \tag{7.10}$$

The quantity $r(N)$ has also been called the *local* or *N-dependent predictability* [38].

To capture the total amount of redundancy per symbol as a measure of memory in an environment, Crutchfield and Feldman [6] define the quantity *Excess Entropy* as

$$E = \lim_{N \to \infty} R(N) \equiv \lim_{N \to \infty} [h(X_1, \ldots, X_N) - N\overline{h}] \tag{7.11}$$

which is the limit of the redundancy in Eq. (7.9). We contrast our results with the excess entropy in later examples.

The entropy and the differential entropy rate are the primary workhorses in agent learning analyses related to reinforcement learning and curiosity learning scenarios [4, 6]. As a result, the definitions of information gain and redundancy from the agent learning literature as presented in this current section are perhaps too expansive and too imprecise in several ways and should be, and can be, refined to allow the observation of additional phenomena.

7.4   Mutual Information in Agent Learning

A recently introduced quantity, *Mutual Information Gain*, is more aligned with Shannon theory and allows a more detailed parsing of what is happening in the learning process [39]. Even though a relative entropy between two probability densities has been called the information gain in the agent learning literature, it can be shown that this definition is just a conditional entropy. As such, the nomenclature, information gain, is misleading.

In terms of information gain, the quantity of interest is the mutual information between the overall sequence and the growing history of the past given by

$$I(X_N; X^{N-1}) = h(X_N) - h(X_N|X^{N-1}) \tag{7.12}$$

The mutual information in Eq. (7.12) is much more intuitive as a measure of information gained as a a function of N and includes the entropy gain from agent learning as a natural component. We can obtain more insight by expanding Eq. (7.12) using the chain rule for mutual information [2] as

$$\begin{aligned} I(X_N; X^{N-1}) &= h(X_N) - h(X_N|X^{N-1}) \\ &= \sum_{k=1}^{N-1} I(X_N; X_k|X_{k-1}, \ldots, X_0) \end{aligned} \tag{7.13}$$

Since $I(X_N; X_{k-1}|X_{k-2}, \ldots, X_0) \geq 0$, we see that $I(X_N; X^{N-1})$ is nondecreasing in N. Therefore, we can characterize the change in mutual information with increasing knowledge of the past history of the sequence as a sum of conditional mutual informations $I(X_N; X_{k-1}|X_{k-2}, \ldots, X_0)$ [39]. We denote $I(X_N; X^{N-1})$ as the *total mutual information gain*.

We can also consider the mutual information between the input sequence X_N and the immediate past values X_{N-M}, $M < N$, which is

$$\begin{aligned} I(X_N; X^{N-M}) &= h(X_N) - h(X_N|X^{N-M}) \\ &= \sum_{k=1}^{M} I(X_N; X_{N-k}|X_{N-k-1}, \ldots, X_{N-M}) \\ &= I(X_N; X_{N-1}|X_{N-2}, \ldots, X_{N-M}) \\ &\quad + \ldots + I(X_N; X_{N-M-1}|X_{N-M}) + I(X_N; X_{N-M}) \end{aligned} \tag{7.14}$$

This expression allows the input block length N to be finite if we need it to be and it also allows the past history M to be finite, which may occur due to having a finite memory for the analyses.

We can say more if the sequence X_k is stationary and Gaussian with $EX_k = 0$, $EX_k X_{k+n} = \rho_n$, and $EX_k^2 = \sigma^2$. Then, we know that

$$h(X_N|X^{M-1}) = \frac{1}{2}\log 2\pi e \, MMSPE(M) \tag{7.15}$$

with $MMSPE(M) = \frac{|\mathcal{R}_{M+1}|}{|\mathcal{R}_M|}$, where the matrices are populated with the ρ_n terms. With stationary and independent X_k, then $h(X_N) = h(X) = \frac{1}{2}\log 2\pi e \sigma^2$, so using Eqs. (7.14) and (9.1), we find that

$$I(X_N; X^{N-M}) = h(X_N) - h(X_N|X^{N-M}) = \frac{1}{2}\log \frac{\sigma^2}{MMSPE(M)} \tag{7.16}$$

This is an important expression for the mutual information gain since knowing the sequence variance and the minimum mean squared prediction error for an Mth order one step ahead predictor, we can evaluate total mutual information gain without having to approximate the probability distributions and the entropies.

Estimation and Entropy Power

8.1 Introduction

Minimum mean squared error estimation. prediction, and smoothing are usually considered to be different measures of performance in relation to mutual information, except in the special case of Gaussian distributions. However, recent results have established a relationship between entropy power and minimum mean squared estimation error quantities. In this chapter, we developed these relationships and show that the change in mutual information obtained by minimum mean squared error filtering, smoothing, and prediction can be expressed as the log ratio of entropy powers. The key result is that the ratios of entropy powers is the ratio of mean squared errors for an important set of probability densities whose differential entropy has a simple form.

8.2 Minimum Mean Squared Error (MMSE) Estimation

In MMSE, the estimation error to be minimized is

$$\epsilon^2 = E(X[k] - \widehat{X}[k|j])^2 \tag{8.1}$$

at time instant k given observations up to and including time instant j. Using the estimation counterpart to Fano's Inequality [2]

$$E(X[k] - \widehat{X}[k|j])^2 \geq \frac{1}{2\pi e} \exp 2[h(X[k]|\widehat{X}[k|j]) \equiv Q_{X[k]|\widehat{X}[k|j]} \tag{8.2}$$

Taking the logarithm of the right side of Eq. (8.2) past the inequality, we obtain

$$h(X[k]|\widehat{X}[k|j]) = \frac{1}{2} \log(2\pi e Q_{X[k]|\widehat{X}[k|j]}) \tag{8.3}$$

© The Author(s), under exclusive license to Springer Nature Switzerland AG 2025
J. D. Gibson, *Information Theoretic Principles for Agent Learning*, Synthesis Lectures on Engineering, Science, and Technology, https://doi.org/10.1007/978-3-031-65388-9_8

Subtracting $h(X[k]|\widehat{X}[k|l])$, $l \neq j$, from the left side of Eq. (8.3) and the corresponding entropy power expression from the right side, we get

$$h(X[k]|\widehat{X}[k|j]) - h(X[k]|\widehat{X}[k|l]) = \frac{1}{2}\log(\frac{Q_{X[k]|\widehat{X}[k|j]}}{Q_{X[k]|\widehat{X}[k|l]}}) \qquad (8.4)$$

Note that the $2\pi e$ divides out in the ratio of entropy powers.

Adding and subtracting $h(X[k])$ from both sides of Eq. (8.4), we can write

$$I(X[k]; \widehat{X}[k|j]) - I(X[k]; \widehat{X}[k|l]) = \frac{1}{2}\log(\frac{Q_{X[k]|\widehat{X}[k|j]}}{Q_{X[k]|\widehat{X}[k|l]}}) \qquad (8.5)$$

Therefore, the difference between the mutual information of $X[k]$ and $\widehat{X}[k|j]$ and the mutual information of $X[k]$ and $\widehat{X}[k|l]$ can be expressed as one half the log ratio of conditional entropy powers. This allows us to characterize Mutual Information Gain or Loss in MMSE filtering, smoothing, and prediction, as we demonstrate in the following.

8.2.1   MMSE Smoothing

We want to estimate a random scalar signal $x[k]$ given the perhaps noisy measurements $z[j]$ for $j > k$, where k is fixed and j is increasing, based on a minimum mean squared error cost function. So, the estimation error to be minimized is

$$\epsilon^2 = E(X[k] - \widehat{X}[k|j])^2 \qquad (8.6)$$

Again, using the estimation counterpart to Fano's Inequality [2]

$$E(X[k] - \widehat{X}[k|j])^2 \geq \frac{1}{2\pi e}\exp 2[h(X[k]|\widehat{X}[k|j]) \equiv Q_{X[k]|\widehat{X}[k|j]} \qquad (8.7)$$

As j increases, the optimal smoothing estimate will not increase the MMSE so

$$Q_{X[k]|\widehat{X}[k|j]} \geq Q_{X[k]|\widehat{X}[k|j+1]} \qquad (8.8)$$

Moving $Q_{X[k]|\widehat{X}[k|j+1]}$ over to the left side of Eq. (8.8) and substituting the definition of entropy power for each produces

$$\frac{Q_{X[k]|\widehat{X}[k|j]}}{Q_{X[k]|\widehat{X}[k|j+1]}} = \exp 2[h(X[k]|\widehat{X}[k|j])] - h(X[k]|\widehat{X}[k|j+1])] \geq 1 \qquad (8.9)$$

Taking logarithms, we see that

$$\frac{1}{2}\log\frac{Q_{X[k]|\widehat{X}[k|j]}}{Q_{X[k]|\widehat{X}[k|j+1]}} = [h(X[k]|\widehat{X}[k|j]) - h(X[k]|\widehat{X}[k|j+1])] \geq 0 \qquad (8.10)$$

Adding and subtracting $h(X[k])$ to the right hand side of Eq. (8.10), yields

$$\frac{1}{2} \log \frac{Q_{X[k]|\widehat{X}[k|j]}}{Q_{X[k]|\widehat{X}[k|j+1]}} = I(X[k]; \widehat{X}[k|j + 1]) - I(X[k]; \widehat{X}[k|j]) \geq 0 \qquad (8.11)$$

Equation (8.11) shows that the mutual information is nondecreasing for increasing $j > k$. Thus we have an expression for the Mutual Information Gain due to smoothing as a function of lookahead in terms of entropy powers.

We can also use Eq. (8.11) to obtain the rate of decrease of the entropy power in terms of the mutual information as

$$Q_{X[k]|\widehat{X}[k|j+1]} = Q_{X[k]|\widehat{X}[k|j]} \exp\left[- 2(I(X[k]; \widehat{X}[k|j + 1]) \right. \\ \left. -I(X[k]; \widehat{X}[k|j])) \right] \qquad (8.12)$$

Here we see that the rate of decrease in the entropy power is exponentially related to the Mutual Information Gain due to smoothing.

8.2.2   MMSE Prediction

We want to predict a random scalar signal $x[k]$ given the perhaps noisy measurements $z[j]$ for $j < k$, where k is fixed and j is decreasing from $k - 1$, based on a minimum mean squared error cost function. So, the prediction error to be minimized is

$$\epsilon^2 = E(X[k] - \widehat{X}[k|j])^2 \qquad (8.13)$$

Again using the estimation counterpart to Fano's Inequality [2]

$$E(X[k] - \widehat{X}[k|j])^2 \geq \frac{1}{2\pi e} \exp 2[h(X[k]|\widehat{X}[k|j]) \equiv Q_{X[k]|\widehat{X}[k|j]} \qquad (8.14)$$

As j decreases, the optimal prediction will increase the minimum mean squared prediction error since the prediction is further ahead, so

$$Q_{X[k]|\widehat{X}[k|j]} \leq Q_{X[k]|\widehat{X}[k|j-1]} \qquad (8.15)$$

Moving $Q_{X[k]|\widehat{X}[k|j]}$ over to the right side of Eq. (8.15) and substituting the definition of entropy power for each produces

$$\frac{Q_{X[k]|\widehat{X}[k|j-1]}}{Q_{X[k]|\widehat{X}[k|j]}} = \exp 2[h(X[k]|\widehat{X}[k|j - 1])] - h(X[k]|\widehat{X}[k|j])] \geq 1 \qquad (8.16)$$

Taking logarithms, we see that

$$\frac{1}{2} \log \frac{Q_{X[k]|\widehat{X}[k|j-1]}}{Q_{X[k]|\widehat{X}[k|j]}} = [h(X[k]|\widehat{X}[k|j-1]) - h(X[k]|\widehat{X}[k|j])] \geq 0 \qquad (8.17)$$

Adding and subtracting $h(X[k])$ to the right hand side of Eq. (8.17), yields

$$\frac{1}{2} \log \frac{Q_{X[k]|\widehat{X}[k|j-1]}}{Q_{X[k]|\widehat{X}[k|j]}} = I(X[k]; \widehat{X}[k|j]) - I(X[k]; \widehat{X}[k|j-1]) \geq 0 \qquad (8.18)$$

This result shows that there is a Mutual Information Loss with further lookahead in prediction and this loss is expressible in terms of a ratio of entropy powers. Equation (8.18) shows that the mutual information is decreasing for decreasing $j < k$, that is, for prediction further ahead, since $I(X[k]; \widehat{X}[k|j]) \geq I(X[k]; \widehat{X}[k|j-1])$. As a result, the observations are becoming less relevant to the variable to be predicted. We can also use Eq. (8.18) to obtain the rate of increase of the entropy power as the prediction is further ahead in terms of the mutual information as

$$Q_{X[k]|\widehat{X}[k|j-1]} = Q_{X[k]|\widehat{X}[k|j]} \exp\left[2(I(X[k]; \widehat{X}[k|j]) \\ -I(X[k]; \widehat{X}[k|j-1]))\right] \qquad (8.19)$$

Thus, the entropy power increase grows exponentially with the Mutual Information Loss corresponding to increasing lookahead in prediction.

8.2.3 MMSE Filtering

We want to estimate a random scalar signal $x[k]$ given the perhaps noisy measurements $z[k]$, based on a minimum mean squared error cost function. So, the estimation error to be minimized is

$$\epsilon^2 = E(X[k] - \widehat{X}[k|k])^2 \qquad (8.20)$$

From the estimation counterpart to Fano's Inequality [2]

$$E(X[k] - \widehat{X}[k|k])^2 \geq \frac{1}{2\pi e} \exp 2[h(X[k]|\widehat{X}[k|k]) \equiv Q_{X[k]|\widehat{X}[k|k]} \qquad (8.21)$$

Dividing $Q_{X[k]|\widehat{X}[k|k]}$ by $Q_{X[k-1]|\widehat{X}[k-1|k-1]}$ and substituting the definition of entropy power for each produces

$$\frac{Q_{X[k]|\widehat{X}[k|k]}}{Q_{X[k-1]|\widehat{X}[k-1|k-1]}} = \exp 2[h(X[k]|\widehat{X}[k|k])]$$
$$-h(X[k-1]|\widehat{X}[k-1|k-1])] \qquad (8.22)$$

Taking logarithms, we see that

$$\frac{1}{2} \log \frac{Q_{X[k]|\widehat{X}[k|k]}}{Q_{X[k-1]|\widehat{X}[k-1|k-1]}} = \left[h(X[k]|\widehat{X}[k|k]) \right.$$
$$\left. - h(X[k-1]|\widehat{X}[k-1|k-1]) \right] \tag{8.23}$$

Adding and subtracting $h(X[k])$ and $h(X[k-1])$ to the right hand side of Eq. (8.23), yields

$$\frac{1}{2} \log \frac{Q_{X[k]|\widehat{X}[k|k]}}{Q_{X[k-1]|\widehat{X}[k-1|k-1]}} = I(X[k-1]; \widehat{X}[k-1|k-1])$$
$$- I(X[k]; \widehat{X}[k|k]) \tag{8.24}$$
$$+ [h(X[k]) - h(X[k-1])]$$

This equation involves the differential entropies of $X[k]$ and $X[k-1]$ unlike prior expressions for smoothing and prediction. This is because the reference points for the two entropy powers are different. However, for certain wide sense stationary processes, we will have simplification as shown in the next section on Entropy Power and MSE.

8.3 Entropy Power and MSE

We know from Sect. 3.3 that the entropy power is the minimum variance that can be associated with a differential entropy $h(X)$. The key insight into relating mean squared error and mutual information comes from considering the (apparently not so special) cases of random variables whose differential entropy has the form

$$h(X) = \frac{1}{2} log[A\sigma_X^2] \tag{8.25}$$

and the log ratio of entropy powers. Therefore, we do not have to explicitly calculate the entropy power since we can use the variance or mean squared error for these distributions in the log ratio of entropy power expressions to find the mutual information gain or loss.

Thus, all of the results in Sect. 8.2 in terms of log ratio of entropy powers can be expressed as ratios of variances or mean squared errors hold for random variables of the form in Eq. (8.25).

In particular, for the smoothing problem, we can rewrite Eq. (8.11) as

$$\frac{1}{2} \log \frac{var(X[k]|\widehat{X}[k|j])}{var(X[k]|\widehat{X}[k|j+1])} = I(X[k]; \widehat{X}[k|j+1])$$
$$- I(X[k]; \widehat{X}[k|j]) \geq 0 \tag{8.26}$$

and the decrease in MSE in terms of the change in mutual information as

$$var(X[k]|\widehat{X}[k|j+1]) = var(X[k]|\widehat{X}[k|j]) \exp \left[-2(I(X[k]; \widehat{X}[k|j+1]) \right.$$
$$\left. - I(X[k]; \widehat{X}[k|j])) \right] \tag{8.27}$$

Here we see that the rate of decrease in the MMSE is exponentially related to the Mutual Information Gain due to smoothing.

Rewriting the results for prediction in terms of variances, we have that Eq. (8.19) becomes

$$
\frac{1}{2} \log \frac{var(X[k]|\widehat{X}[k|j-1])}{var(X[k]|\widehat{X}[k|j])} = I(X[k]; \widehat{X}[k|j])
$$

$$
-I(X[k]; \widehat{X}[k|j-1]) \geq 0
$$

(8.28)

and that the growth in MMSE with increasing lookahead is

$$
var(X[k]|\widehat{X}[k|j-1]) = var(X[k]|\widehat{X}[k|j]) \exp\left[2(I(X[k]; \widehat{X}[k|j])\right.
$$

$$
\left. -I(X[k]; \widehat{X}[k|j-1]))\right]
$$

(8.29)

Thus, as lookahead in prediction is increased, the conditional error variance grows exponentially.

For the filtering problem, we have the two differential entropies, $h(X[k])$ and $h(X[k-1])$ in Eq. (8.24) in addition to the mutual information expressions. However, for wide sense stationary random processes with differential entropies of the form shown in Eq. (8.25), the two variances are equal so $var(X[k]) = var(X[k-1])$ so the difference in the two differential entropies is zero. This simplifies Eq. (8.24) to

$$
\frac{1}{2} \log \frac{var(X[k]|\widehat{X}[k|k])}{var(X[k-1]|\widehat{X}[k-1|k-1])} = I(X[k-1]; \widehat{X}[k-1|k-1])
$$

$$
-I(X[k]; \widehat{X}[k|k])
$$

$$
\leq 0
$$

(8.30)

which if the error variance in monotonically nonincreasing is less than or equal to zero as shown. Rewriting this last result in terms of increasing mutual information, we have

$$
\frac{1}{2} \log \frac{var(X[k-1]|\widehat{X}[k-1|k-1])}{var(X[k]|\widehat{X}[k|k])} = I(X[k]; \widehat{X}[k|k])
$$

$$
-I(X[k-1]; \widehat{X}[k-1|k-1])
$$

$$
\geq 0
$$

(8.31)

It is important to note the power of the expressions in this section. We are able to obtain the mutual information gain or loss by using the variances of MMSE estimators. There is no need to utilize techniques to approximately compute mutual informations, which are fraught with difficulties. See Hudson [40].

Time Series Analyses

9

9.1 Introduction

Statistical time series analysis and time series models for physical systems have a long and accomplished history. It is important when trying to evaluate what an agent has learned or can learn from time series observations that classical results are understood and exploited. Of course, many researchers in statistical data science are well schooled in the methods and practices of time series analysis.

In this chapter, we explore some of the classical results and develop information theoretic representations relevant to these classical time series analysis results. Particular focus is placed on autoregressive (AR) models since they have played a fundamental role in describing many physical processes. We also explore an information theoretic decomposition of one of the major successes of AR modeling, Code Excited Linear Prediction for the compression of speech signals. This topic appears particularly relevant due to the great success of large language models and deep learning in speech recognition.

Many other applications of deep learning to speech processing are also being explored. In fact, machine learning research for speech coding has been pursued in recent years, but has yet to prove competitive with code excited linear prediction established over the past 25 years.

9.2 Stationary and Gaussian

We can say more if the sequence X_k is stationary and Gaussian with $E X_k = 0$, $E X_k X_{k+n} = \rho_n$, and $E X_k^2 = \sigma^2$. Then, we know that

$$h(X_N | X^{M-1}) = \frac{1}{2} \log 2\pi e MMSPE(M) \tag{9.1}$$

© The Author(s), under exclusive license to Springer Nature Switzerland AG 2025   49
J. D. Gibson, *Information Theoretic Principles for Agent Learning*, Synthesis Lectures on Engineering, Science, and Technology, https://doi.org/10.1007/978-3-031-65388-9_9

with $MMSPE(M) = \frac{|\mathcal{R}_{M+1}|}{|\mathcal{R}_M|}$, where the matrices are populated with the ρ_n terms. With stationary X_k, then $h(X_N) = h(X) = \frac{1}{2} \log 2\pi e \sigma^2$, so using Eqs. (7.4) and (7.14), we find that

$$I(X_N; X^{N-M}) = h(X_N) - h(X_N|X^{N-M}) = \frac{1}{2} \log \frac{\sigma^2}{MMSPE(M)} \quad (9.2)$$

This is an important expression for the mutual information gain since knowing the sequence variance and the minimum mean squared prediction error for an Mth order predictor, we can evaluate total mutual information gain without having to approximate the probability distributions and the entropies.

The utility of the mutual information gain expressions in Eqs. (7.13) and (7.14) becomes even more evident under the Gaussian assumption since the conditional mutual information terms become

$$\begin{aligned} I(X_N; X_k|X_{k-1}, \ldots, X_{N-M}) &= h(X_N|X_{k-1}, \ldots, X_{N-M}) \\ &\quad - h(X_N|X_k, X_{k-1}, \ldots, X_{N-M}) \\ &= \frac{1}{2} \log \frac{\sigma^2_{e(k-1)}}{\sigma^2_{ek}} \end{aligned} \quad (9.3)$$

Then we have for Eq. (7.14)

$$\begin{aligned} I(X_N; X^{N-M}) &= h(X_N) - h(X_N|X^{N-M}) \\ &= \frac{1}{2} [\log \frac{\sigma^2}{\sigma^2_{e1}} + \log \frac{\sigma^2_{e1}}{\sigma^2_{e2}} \cdots + \log \frac{\sigma^2_{e(N-M-1)}}{\sigma^2_{e(N-M)}}] \end{aligned} \quad (9.4)$$

We know that the minimum mean squared prediction error is nonincreasing $\sigma^2_{e(n-1)} \geq \sigma^2_{e(n)}$, so each term in the sum in Eq. (9.4) is greater than or equal to zero, as must be true since it is a mutual information.

We see from Eqs. (7.14), (9.2), and (9.4) that the mutual information gain gives us a quantitative indicator in bits/symbol of the linear redundancy being captured or modeled. This is a new and useful indicator of structure or memory being separated from randomness.

9.2.1 A Distribution Free Information Measure

If we compare the mutual information in Eq. (9.2) with the entropy gain expression in Eq. (7.4), the scaling factor for the Gaussian density has been divided out and is not present in Eq. (9.2). This lack of scaling is important when interpreting the mutual information gain since it is no longer dependent on the underlying distribution that would create a bias term.

Note that the prior definition of information gain in agent learning in Eq. (7.5) is actually an entropy gain so the scaling factor is present. The new quantity, total mutual information gain, therefore, has a distribution free property not satisfied by entropy gain. In fact, for continuous random variables, the differential entropy can be changed by a linear transformation but the mutual information cannot [2].

9.3 Autoregressive Modeling

An autoregessive (AR) process is given by

$$x(k) = \sum_{i=1}^{M} a_i x(k-i) + w(k) \tag{9.5}$$

where the a_i, $i = 1, 2, \ldots M$ are called autoregressive parameters and $w(k)$ is the excitation sequence. Let us assume that the sequence being analyzed is a stationary, purely autoregressive sequence of order M and the excitation term $w(k)$ has the possibly nonuniform probability density function $p_W(w)$ with variance σ^2.

In general, the distance from randomness consists of two components, the linear redundancy due to the predictive component and the nonlinear redundancy due to the distribution of the excitation [39].

If we know the true autoregressive parameters and the correct AR model order for a sequence, then the linear redundancy can be removed by operating on the given sequence so that the remaining distance from randomness is the nonlinear redundancy only. However, in most learning and modeling problems, even if we are willing to assume that the sequence being observed is autoregressive, the true AR model order is not known. The following example explores these ideas.

Example: Learning and Modeling an AR Sequence

A zero mean unit variance purely AR(10) Gaussian sequence is given by Eq. (9.5) with coefficients $a_1 = 2.0965$, $a_2 = -2.6235$, $a_3 = 1.4123$, $a_4 = -0.8282$, $a_5 = 0.5066$, $a_6 = -0.1511$, $a_7 = -0.7505$, $a_8 = 1.1628$, $a_9 = -0.7748$, $a_{10} = 0.1906$, where the sequence $w(k)$ is Gaussian with zero mean and variance σ_W^2. (Note that these autoregressive parameters, a_i, $i = 1, 2, \ldots M$, were obtained by processing a frame of speech sampled at 8000 samples/sec to calculate the autocorrelation terms and then using standard techniques to solve the linear simultaneous equations for the coefficients) [8]. Table 9.1 shows the incremental mutual information gain and the total mutual information gain as the predictor order M is increased.

We observe that the MMSPE ($\sigma_{e(M)}^2$) is decreasing monotonically but not so for the incremental mutual information gain, which increases in going from a 1st order predictor to a 2nd order predictor and also in going from a 3rd order predictor to an $M = 4th$ order predictor and further when the predictor order goes from $M = 8$ to $M = 9$. Perhaps this hints at why mean squared error is thought not to be a reliable indicator of performance in learning applications.

Table 9.1 Incremental and total mutual information gain as the predictor order is increased: zero mean, unit variance Gaussian AR(10) sequence

M	$\sigma^2_{e(M)}$	$I(X_N; X_k \mid X_{k-1}, \ldots, X_{N-M})$	$I(X_N; X^{N-M})$
0	1.0	0 bits/symbol	0 bits/symbol
1	0.3111	0.842	0.842
2	0.0667	1.11	1.952
3	0.0587	0.092	2.044
4	0.0385	0.304	2.348
5	0.0375	0.019	2.367
6	0.0342	0.065	2.432
7	0.0308	0.069	2.501
8	0.0308	0.0	2.501
9	0.0261	0.12	2.621
10	0.0251	0.026	2.647
0–10	0.0251	2.647	2.647

However, there is an even tighter connection between these increases in mutual information gain and the physical process inherent in the autoregressive model with the given coefficients. The frequency response corresponding to the AR model in Eq. (9.5) and the given coefficients is plotted in Fig. 9.1. There are three major peaks evident in the spectrum, but certainly the relative magnitudes of the peaks are quite different. As noted from Table 9.1, there are jumps in the incremental mutual information gain was the predictor order changes from 0 to 1, from 1 to 2, from 3 to 4, and from 8 to 9. There is a general rule that to represent a peak in a spectral envelope requires two model coefficients, which when translated to the frequency domain provide the location of the spectral peak and the bandwidth of that peak.

So, the increase in predictor order results in capturing the successive peaks in the spectral envelope. If we were to plot the spectra as the predictor order is increased from 0 to 10, this evolution would be clearer with the substantial jump in incremental mutual information gain in going from 0 to 1 showing a magnitude at low frequencies and rough location of the peak but not the bandwidth (not an isolated peak itself). Further discussion of these ideas are more properly in the context of time series analysis or linear prediction of speech [41] than in the present development of this example; however, it is evident that the incremental mutual information gain indicates significant physical changes in the underlying sequence that, while present in the changes in mean squared prediction error, they are not highlighted as with the incremental mutual information gain.

The total mutual information gain of 2.647 bits/symbol is the gain that comes from the linear redundancy in the AR(10) sequence, and the remaining redundancy is the nonlinear redundancy. If this sequence is modeled with a $M = $ 2nd order predictor, that is, if the AR(10)

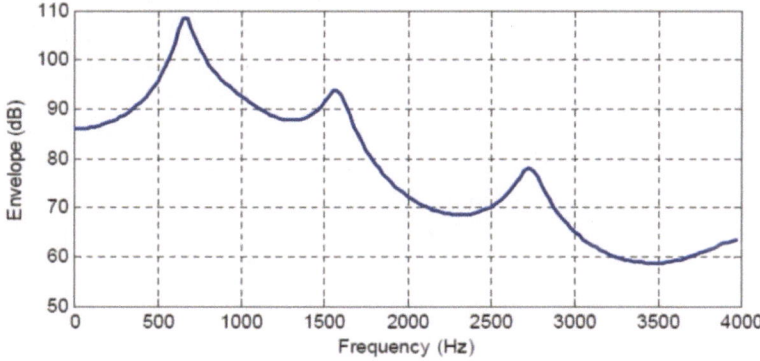

Fig. 9.1 AR model frequency response

sequence is modeled as an AR(2) sequence, we would conclude that the mutual information gain or linear redundancy of such a sequence was only 1.952 bits/symbol with $\sigma_W^2 = \sigma_{e(2)}^2 = 0.0667$.

9.4   Code-Excited Linear Prediction (CELP)

Block diagrams of a code-excited linear prediction (CELP) encoder and decoder are shown in Figs. 9.2 and 9.3, respectively [8, 42].

We provide a brief description of the various blocks in Figs. 9.2 and 9.3 to begin. The CELP encoder is an implementation of the Analysis-by-Synthesis (AbS) paradigm [42]. CELP, like most speech codecs in the last 45 years, is based on the linear prediction model for speech, wherein the speech is modeled as

$$s(k) = \sum_{i=1}^{N} a_i s(k-i) + Gw(k), \tag{9.6}$$

where we see that the current speech sample at time instant k is represented as a weighted linear combination of N prior speech samples plus an excitation term at the current time instant. The weights, $a_i, i = 1, 2, ..., N$, are called the linear prediction coefficients. The synthesis filter in Fig. 9.2 has the form of this linear combination of past outputs and the fixed and adaptive codebooks model the excitation, $w(k)$. The LP analysis block calculates the linear prediction coefficients, and we see that the block also quantizes the coefficients so that encoder and decoder use exactly the same coefficients.

The adaptive codebook is used to capture the long-term memory due to the speaker pitch and the fixed codebook is selected to be an algebraic codebook, which has mostly zero values and only a relatively few nonzero pulses. The pitch analysis block calculates the adaptive

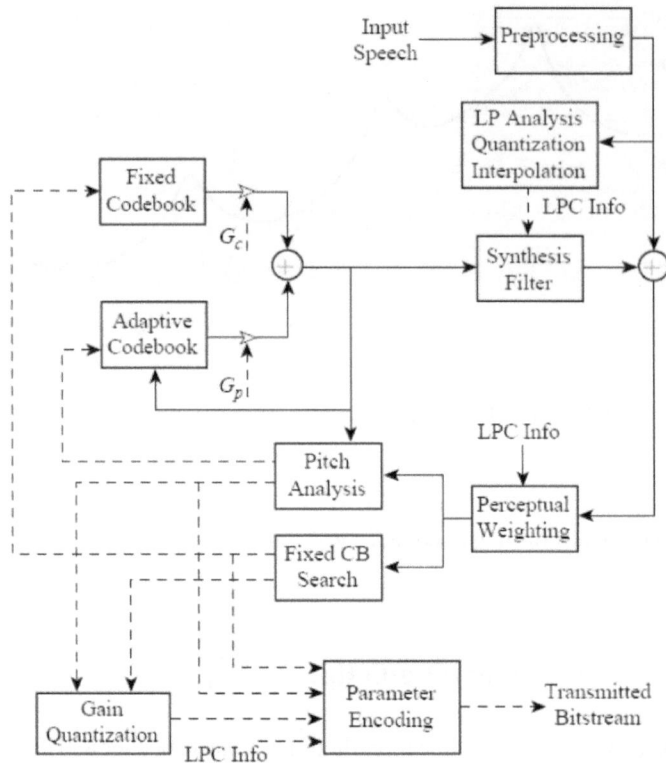

Fig. 9.2 Code-excited linear prediction (CELP) encoder

codebook long-term memory. The process is AbS in that for a block of (say) M input speech samples, the linear prediction coefficients and long-term memory are calculated and a perceptual weighting filter is constructed using the linear prediction coefficients. Then, for every length M sequence (codevector) in the fixed codebook, (say) there are L code vectors in the fixed codebook, a synthesized sequence of speech samples are produced. This is the fixed codebook search block. The best codevector out of the L in the fixed codebook in terms of the length M synthesized sequence that best matches the input block of length M based on minimizing the perceptually weighted squared error is chosen and transmitted to the CELP decoder along with the long-term memory, the predictor coefficients, and the codebook gains. These operations are represented by the parameter encoding block in Fig. 9.2 [42].

The CELP decoder uses these parameters to synthesize the block of M reconstructed speech samples presented to the listener as shown in Fig. 9.3. There is also post-processing, which is not shown in the figure.

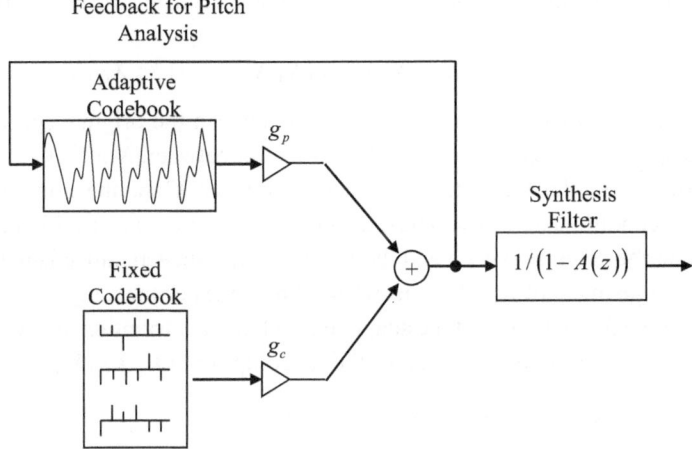

Fig. 9.3 CELP decoder

The quality and intelligibility of the synthesized speech is often determined by listening tests that produce mean opinion scores (MOS), which for narrowband speech vary from 1 up to 5. A well-known codec such as G.711 is usually included to provide an anchor score value with respect to which other narrowband codecs can be evaluated [41].

It would be helpful to be able to associate a separate contribution to the overall performance by each of the main components in Fig. 9.3, namely the fixed codebook, the adaptive codebook, and the synthesis filter. Signal to quantization noise ratio (SNR) is often used for speech waveform codecs, but CELP does not attempt to follow the speech waveform, so SNR is not applicable. One characteristic of CELP codecs that is well known is that those speech attributes not captured by the short-term predictor must be accounted for, as best as possible, by the excitation codebooks, but an objectively meaningful measure of the individual component contributions is yet to be advanced.

In the next section, we propose a decomposition in terms of the mutual information and conditional mutual information with respect to the input speech provided by each component in the CELP structure, that appears particularly useful and interesting for capturing the performance and the trade-offs involved.

9.5   A Mutual Information Decomposition

Gibson [5] proposed the following decomposition of the several contributions to the synthesized speech by the CELP codec components. In particular, letting X represent a frame of input speech samples, we define X_R, X_N, and X_C as the reconstructed speech, the prediction component, and the combined fixed and adaptive codebook components, respectively. Then we can write the mutual information between the input speech and the reconstructed speech as

$$I(X; X_R) = I(X; X_N, X_C) = I(X; X_N) + I(X; X_C|X_N). \tag{9.7}$$

This expression states that the mutual information between the original speech X and the reconstructed speech X_R equals the mutual information between X and X_N, the Nth order linear prediction of X, plus the mutual information between X and the combined codebook excitations X_C conditioned on X_N. Thus, to achieve or maintain a specified mutual information between the original speech and the reconstructed speech, any change in X_N must be offset by an adjustment of X_C. This fits what is known experimentally and was alluded to earlier. If we define X_A to represent the adaptive codebook contribution and X_F to represent the fixed codebook contribution, we can further decompose $I(X; X_C|X_N)$ as

$$\begin{aligned} I(X; X_C|X_N) &= I(X; X_A, X_F|X_N) \\ &= I(X; X_A|X_N) + I(X; X_F|X_N, X_A), \end{aligned} \tag{9.8}$$

where we have used the chain rule for mutual information [2]. The expressions in Eqs. (9.7) and (9.8) correspond to the analysis chain illustrated in Fig. 9.4, so $X \to X_N \to X_A \to X_F$. $A(z)$ in Fig. 9.4 represents the short-term prediction component as in Eq. (9.6), and $P(z)$ is the long-term predictor.

We can also write a chain rule mutual information expression for the synthesis chain in Fig. 9.5 as

$$\begin{aligned} I(X; X_R) &= I(X; X_C) + I(X; X_N|X_C) \\ &= I(X; X_A, X_F) + I(X; X_N|X_C) \\ &= I(X; X_F) + I(X; X_A|X_F) + I(X; X_N|X_A, X_F)), \end{aligned} \tag{9.9}$$

so $X_F \to X_A \to X_N \to X$.

Notice that we are not attempting to model the mutual informations in the CELP encoder and decoder of Figs. 9.2 and 9.3 directly; we are effectively creating analysis and synthesis Markov chains that use the choices for X_F, X_A, X_N produced by the CELP encoder in the

Fig. 9.4 Analysis chain

Fig. 9.5 Synthesis chain

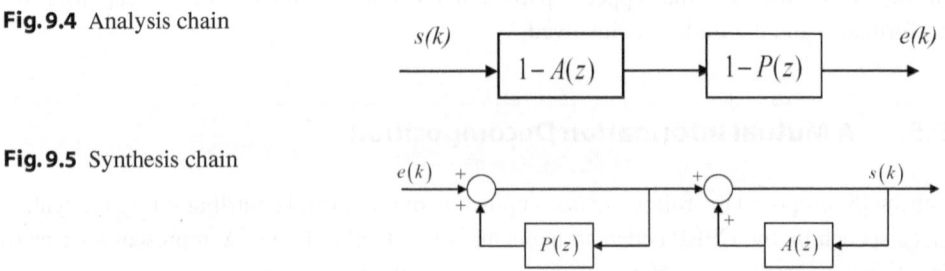

original analysis-by-synthesis CELP structure prior to the adoption of the adaptive codebook approximation.

While these expressions in Eqs. (9.7)–(9.9) are interesting, the challenge that remains is to characterize each of these mutual informations without actually having to calculate them directly from data, which is a difficult problem in and of itself [43].

An interesting quantity introduced and analyzed by Gibson in a series of papers is the log ratio of entropy powers [34–36]. Specifically, the log ratio of entropy powers is related to the difference in mutual information, and further, in many cases, the entropy powers can be replaced with the minimum mean squared prediction error (MMSPE) in the ratio. Using the MMSPE, the difference in mutual informations can be easily calculated.

Information Bottleneck Principle 10

10.1 Introduction

In information theory, it is well known that for a Markov Chain, $X_1 \to X_2 \to X_3$, if the number of states in X_1 is k and the number of states in X_2 is m with $m < k$, then $I(X_1; X_3) \leq \log m$. Thus, a bottleneck limits the amount of mutual information between the input to the Markov chain a random variable later than the bottleneck. With this in mind, a theoretical framework for analyzing deep neural networks called the Information Bottleneck has been proposed, where the desire is to have the simplest possible representation that captures all of the relevant information. The simple representation is what produces the information bottleneck.

10.2 The Problem Formulation

We begin by stating the formulation and then later address some claims about the Information Bottleneck Method in the literature. With X representing the input layer to a neural network (NN), then during training the desired output Y is observed probabilistically during the training phase through the joint distribution $p(X, Y)$. The mutual information $I(X; Y)$ indicates the relevant information that an input X has about the desired output Y.

After training, an input X is observed and is processed through the network to the output layer that is the predicted desired output $\widehat{Y}$. The optimal representation of the input X that captures the relevant information is a compressed version of X, denoted by $\widehat{X}$. Hopefully, it is the simplest compression of the input X that contains the relevant information about Y per the mutual information $I(X; Y)$. All of this leads to the Markov Chain

$$Y \to X \to \widehat{X}, \tag{10.1}$$

© The Author(s), under exclusive license to Springer Nature Switzerland AG 2025   59
J. D. Gibson, *Information Theoretic Principles for Agent Learning*, Synthesis Lectures on Engineering, Science, and Technology, https://doi.org/10.1007/978-3-031-65388-9_10

where $\widehat{X}$ is the representation generated via the internal stages of the NN. The Markov Chain in Eq. (10.1) is unusual in that it involves the desired output Y only available during training and not the predicted desired output of the NN $\widehat{Y}$.

Based on the Markov Chain in Eq. (10.1), the optimization problem addressed by the information bottleneck method can be defined as desiring to minimize the mutual information $I(X; \widehat{X})$ subject to the constraint $I(\widehat{X}; Y)$. The interpretation of this problem is that $I(X; \widehat{X})$ captures the complexity of the representation through the rate associated with the compression and $I(\widehat{X}; Y)$ expresses the amount of the information preserved about the trained variable by $\widehat{X}$, the compressed variable.

The predicted variable $\widehat{Y}$ is extracted from $\widehat{X}$ so that based on the Markov Chain in Eq. (10.1), $I(X; Y) \geq I(Y; \widehat{Y})$, with equality iff $\widehat{X}$ is a sufficient statistic for X.

10.3   The Optimization Problem

The optimization problem then can be expressed as minimizing the Lagrangian

$$J(p(x|\hat{x})) = I(X; \widehat{X}) - \lambda I(\widehat{X}; Y) \tag{10.2}$$

where by adjusting the Lagrange parameter λ there is a tradeoff between the rate of the representation $\widehat{X}$ and the preserved information about Y from training.

Solving the optimization problem yields the following probability mass functions

$$p(\hat{x}|x) = \frac{p(\hat{x})}{\mu(x; \lambda)} \exp\left(-\lambda D(p(y|x)\|p(y|\hat{x}))\right) \tag{10.3}$$

with

$$p(y|\hat{x}) = \sum_x p(y|x)p(x|\hat{x}) \tag{10.4}$$

and

$$p(\hat{x}) = \sum_x p(x)p(\hat{x}|x) \tag{10.5}$$

The quantity $\mu(x; \lambda)$ is chosen to impose the fact that $p(\hat{x}|x)$ must sum to 1. There is a great similarity between these equations and the Lagrangian minimization problem that characterizes the rate distortion function with a distortion measure $d(x, \hat{x})$ in classical rate distortion theory. For the information bottleneck method, no distortion measure is chosen and the relative entropy $D(p(y|x)\|p(y|\hat{x}))$ fulfills that role.

This result is interpreted as an optimization problem that depends on a nonfixed distortion measure

$$d_{IB}(x, \hat{x}) = D(p(y|x)\|p(y|\hat{x})) \tag{10.6}$$

This is an interesting interpretation, particularly when the similarity to successive refinement of information is invoked. In classical rate distortion theoretic successive refinement, the distortion measure can change at every stage but the successive stages must identically cover the preceding stage average distortion.

Channel Capacity

11

11.1   Introduction

While channel capacity is inextricably tied to the communications problem, if properly understood, it can prove useful in gaining insights into the transfer of information regarding sequences through systems that can be modeled by conditional probabilities. In the communications problem, the conditional probabilities constitute the channel model.

Then, the basic problem is how to choose the input distribution to such models so as to maximize the rate of information passed through the system while maintaining an asymptotically small probability of error in reproducing the information at the output. When stated this way, the utility of channel capacity in agent learning analyses is much more direct.

11.2   The Definition of Channel Capacity

In the classical Shannon channel capacity theorem, we are interested in the transmission of information over a noisy channel. More specifically, we would like to address the question: Given the characterization of a communications channel, more generally, a system model, what is the maximum bit rate that can be sent over this channel with negligibly small error probability? We find that the mutual information between the channel input and output random variables plays an important role in providing the answer to this question. Since mutual information is often used in analyses of agent learning, it is worthwhile to understand the structure and development of channel capacity to assess if it might be of interest in characterizing the performance of agent learning.

To introduce the approach and the fundamental ideas, we begin by considering *channels or systems* that have finite input and output alphabets and for which the output letter at any given time depends only on the channel input letter at the same time instant. Therefore, for an

© The Author(s), under exclusive license to Springer Nature Switzerland AG 2025   63
J. D. Gibson, *Information Theoretic Principles for Agent Learning*, Synthesis Lectures on Engineering, Science, and Technology, https://doi.org/10.1007/978-3-031-65388-9_11

input alphabet $W = \{1, 2, \ldots, M\}$ with probability assignment $P_W(w)$, $w = 1, 2, \ldots, M$, and output alphabet $X = \{1, 2, \ldots, N\}$ the system is described by the *transition probabilities* $P_{X|W}(x|w)$, $w = 1, 2, \ldots, M$, and $x = 1, 2, \ldots, N$. From these quantities we can calculate the mutual information between the channel or system input W and output X. Then, the Capacity of a channel or system with input W and output X is defined as

$$C \overset{\triangle}{=} \max_{\text{all } P_W(\cdot)} I(W; X)$$

$$= \max_{\text{all } P_W(\cdot)} \sum_{w=1}^{M} \sum_{x=1}^{N} P_{WX}(w, x) \log \frac{P_{WX}(w, x)}{P_W(w) P_X(x)} , \tag{11.1}$$

where the maximum is taken over all channel (system) input probability assignments.

We note that $I(W; X)$ is a function of the input probabilities and the transition probabilities, whereas the channel capacity C is a function of the input probabilities only. Once we have selected a physically appropriate model for the channel (system), we assume that we have no control over the transition probabilities, since they are determined by the channel/system behavior as represented by the model. In words, Eq. (11.1) says that the capacity of a DMC (discrete memoryless channel) is the largest mutual information between the input and output that can be transmitted over the channel in one use.

The physical significance of the channel capacity expression is illustrated by the following theorem:

Shannon's Second Theorem (Channel Coding Theorem) [1, 2, 7] Capacity C expressed as

$$C = \max_{p(x)} I(X; Y) . \tag{11.2}$$

is the maximum rate that we can send information over the channel and recover the information at the channel output with vanishingly small error probability.

We present very few detailed proofs in this book, but later in the current chapter we provide an outline of a proof because of the physical insights that can be gained from the basic approach. It is emphasized again that the mathematical expression for capacity in terms of mutual information only has physical meaning because of this theorem. The physical significance of the mathematical capacity expression that results from the maximization depends on the accuracy of the channel/system model in representing the actual physical channel/system.

Therefore, for the chosen channel/system model, Channel Capacity is a fundamental limit on the maximum data rate that can be transmitted reliably over the channel. As a consequence, Channel Capacity is a physically significant quantity and has a myriad of uses. One such use is to allow a researcher to determine how close current or proposed approaches for information transmission over the channel/system are to the fundamental limit on the reliable transmitted rate. The principal implication being that if one is already operating close to channel/system capacity, perhaps additional effort or added complexity to further improve performance is not warranted. Insights like this would be especially useful in learning applications.

To calculate the channel capacity, it is necessary, as indicated by Eq. (11.1), to perform a maximization over M variables, the $P_W(w)$, subject to the constraints that $P_W(w) \geq 0$ for all w and $\sum_{w=1}^{M} P_W(w) = 1$. In general, this is a difficult task. Later in the chapter, we present approaches for obtaining channel capacity expressions for different channel models.

11.3   Properties of Channel Capacity

There are several key properties of mutual information and channel capacity that are helpful in evaluating channel capacity for a chosen channel model or for performing analyses that reveal fundamental insights for communication systems. A few of these are presented now:

1. $C \geq 0$, since $I(X; Y) \geq 0$.
2. $C \leq \log |X|$ since $C = \max_{p(x)} I(X; Y) \leq \max H(X) = \log |X|$.
3. $C \leq \log |Y|$.
4. $I(X; Y)$ is a continuous function of $p(x)$.
5. $I(X; Y)$ is a concave function of $p(x)$.

Since $I(X; Y)$ is a concave function over a closed convex set, a local maximum is a global maximum. From Properties 2 and 3, the maximum is finite and we can use maximum rather than supremum (sup).

11.4   Calculating Capacity for Discrete Memoryless Channels

It is often difficult to know how to begin to find capacity once a meaningful channel/system model has been chosen. One way to simplify the calculation of channel capacity is to exploit structure in the channel/system model. An important example of such structure is symmetry in the channel probability transition matrix (PTM). The PTM is defined as the matrix of channel transition probabilities where the inputs are rows and the outputs are columns.

Letting $X^n = (x_1, x_2, \ldots, x_n)$ be an input sequence and $Y^n = (y_1, y_2, \ldots, y_n)$ be an output sequence, the nth use of a DMC is described by

$$p\left(Y^n|X^n\right) = \prod_{i=1}^{n} p\left(y_i|x_i\right) . \tag{11.3}$$

Necessary and sufficient conditions for a set of input probabilities $P_W(w)$, $w = 1, 2, \ldots, M$, to achieve capacity is for

$$I\left(W = w; X\right) = C \qquad \text{for all } w \text{ with } P_W\left(w\right) > 0$$

and

$$I\left(W=w;X\right)\leq C \qquad \text{for all } w \text{ with } P_W\left(w\right)=0\ ,$$

for some number C, where

$$I\left(W=w;X\right)=\sum_{x=1}^{N} P_{X|W}\left(x|w\right)\log\frac{P_{X|W}(x|w)}{\sum_{w=1}^{M} P_W(w)P_{X|W}(x|w)}\ .$$

The number C is the channel capacity.

The main use of this theorem is to check the validity of some hypothesized set of input probabilities; that is, guess and verify. Thus, for a binary memoryless channel with symmetric transition probabilities, we might guess by symmetry that $P_W(0) = P_W(1) = \frac{1}{2}$ achieves capacity and then check that the preceding conditions hold.

11.5   The Channel Coding Theorem

Shannon introduced the idea of channel capacity in 1948, when he defined capacity as the maximum rate that one can transmit over a channel with an arbitrarily small error probability. Conceptually, this would seem impossible; that is, to achieve an arbitrarily small error probability for a noisy channel. However, Shannon based his development of capacity on some extraordinarily innovative new insights.

First, he was trying to understand what is possible, separate from how to do it. Specifically, he was after an existence proof; he wanted to show that there *was* a codebook (set of input probabilities) that achieved an arbitrarily small error probability. Second, he was willing to trade delay for performance. Today, we would equate increased delay with increased signal processing complexity. Third, Shannon considered random codebooks, including all possible good and bad codebooks, and fourth, he allowed asymptotically long codewords. He then considered the probability of error averaged over all possible codebooks, good and bad. This approach combined with the arbitrarily long codewords allowed him to show that the average error probability over all codebooks has an exponentially decreasing error probability. Finally, he observed that if this last fact were true, then there must be at least one good codebook in the set. This collection of ideas was nothing short of revolutionary. Today, we might call Shannon's proof of channel capacity a disruptive technology, and that would be an understatement.

Given the list of Shannon's insights and the basic ideas behind channel capacity, it is likely still not evident to the reader how to go about proving the channel coding theorem nor why the mutual information turns out to be an important quantity. In the following, we first outline the required proof and align each step with Shannon's assumptions, and then we explain how mutual information comes to play such an important role in the expression for capacity.

Shannon's Second Theorem (Channel Coding Theorem) [1, 2, 7]   Capacity C expressed as

$$C = \max_{p(x)} I(X; Y). \tag{11.4}$$

is the maximum rate that we can send information over the channel and recover the information at the channel output with vanishingly small error probability. More specifically, if the transmission rate $R < C$, then there exists a sequence of length-n codes with error probability $P_r^{(n)}(E)$ that approaches 0 as n increases, and conversely, for any sequence of codes with $P_r^{(n)}(E) \to 0$ with increasing n, then $R \leq C$.

Outline of the Proof that rates $R < C$ are achievable: If we have M messages to communicate length n sequences, then the rate R of the code is related to M by $M = 2^{nR}$, and thus there are 2^{nR} codewords.

For a fixed $p(x)$, we independently generate length n codewords according to the distribution

$$p(x^n) = \prod_{i=1}^{n} p(x_i) . \tag{11.5}$$

so that the probability of a particular code C is

$$P_r(C) = \prod_{w=1}^{2^{nR}} \prod_{i=1}^{n} p(x_i(w)) . \tag{11.6}$$

where each element is generated iid according to $p(x)$.

With these steps as setup, the communication system operates as follows: A random code C is designed (generated) by (11.6) according to $p(x)$, and the code is known to both the transmitter and receiver. The transmitter and receiver are also assumed to know the channel transition matrix $p(y|x)$.

A message W to be sent is chosen according to a uniform distribution,

$$P_r(W = w) = 2^{-nR}, \qquad w = 1, 2, \ldots, 2^{nR} . \tag{11.7}$$

and the wth codeword $X^n(w)$ is sent over the channel.

A sequence Y^n arrives at the receiver according to the distribution

$$P(y^n | x^n(w)) = \prod_{i=1}^{n} p(y_i | x_i(w)) . \tag{11.8}$$

The receiver decodes the received sequence as $\hat{W}$, an estimate of the input message W. There is a decoding error if $\hat{W} \neq W$.

Let E be the event $\{\hat{W} \neq W\}$.

We can express the average probability of error, averaged over all codewords in the codebook, and averaged over all codebooks, as

$$P_r(E) = \sum_C P(C) P_e^{(n)}(C) \tag{11.9}$$

$$= \sum_C P(C) \frac{1}{2^{nR}} \sum_{w=1}^{2^{nR}} p_w(C) \tag{11.10}$$

$$= \frac{1}{2^{nR}} \sum_{w=1}^{2^{nR}} \sum_C P(C) p_w(C) , \tag{11.11}$$

where $P_e^{(n)}(C)$ is defined for whatever decoding rule is implemented (we will choose a decoding rule and elaborate on this statement shortly). It is this decoding error that we must prove is exponentially small with increasing n.

So, the next question that we must address is what is the form of the decoder? The most common proofs of the channel coding theorem use either maximum likelihood decoding or decoding using jointly typical sequences. Maximum likelihood (ML) decoding is a structure that is very familiar to communications systems designers, so it would appear that ML decoding would be preferable. However, it turns out that the analysis using ML decoding is more difficult mathematically and nonintuitive.

Decoding using jointly typical sequences is likely a new concept and decoding using joint typicality is more difficult to envision as a physical receiver structure, but the proof of the channel coding theorem is more straightforward and produces a more intuitive result than the ML decoder approach. Furthermore, although decoding using joint typicality is suboptimal, it allows us to show that the average error probability asymptotically approaches zero and that is all that we need.

In the next section, we introduce the definition of jointly typical sequences and the joint Asymptotic Equipartition Property (AEP), and show how these can lead to the desired result.

11.6   Decoding and Jointly Typical Sequences

Given a channel output Y^n, we need to decide what input sequence was transmitted. Each input sequence is represented for transmission over the channel by a codeword, so if the message to be transmitted has index i, then the transmitted codeword is $X^n(i)$. Using joint typicality, we decode a channel output Y^n as the ith index if the codeword $X^n(i)$ is "jointly typical" with the received signal Y^n. To analyze the error probability, we define the concept of joint typicality and find the probability of joint typicality when Y^n is due to the input codeword $X^n(i)$ and the probability of Y^n occurring when $X^n(i)$ is not the transmitted codeword.

The proofs are left to the references. However, we can now provide some motivation for the role of mutual information in the calculation of channel capacity.

From the joint AEP, we see that the probability of the two statistically independent sequences (X^n, Y^n), with marginal distributions the same as the transmitted and received codewords, being jointly typical at the receiver is about $2^{-n(I(X;Y))}$ and so we see the appearance of the mutual information in conjunction with jointly typical decoding.

Without going into further mathematical details concerning joint typicality, we can motivate channel capacity as follows. The following development also provides a simple analysis tool for learning agent analyses.

Therefore, the total number of possible (typical) Y sequences is $\cong 2^{nH(Y)}$. For each (typical) input n-sequence, there are approximately $2^{nH(Y|X)}$ possible Y sequences at the receiver, all of them equally likely. For good decoding, we need to operate such that no two X sequences produce the same Y output sequence, or we cannot decode correctly.

Thus, the set of $2^{nH(Y)}$ sequences is to be divided into sets of size $2^{nH(Y|X)}$ corresponding to the different input sequences. Dividing, we see that the total number of disjoint sets is less than or equal to

$$2^{n(H(Y)-H(Y|X))} = 2^{nI(X;Y)} .$$

Hence, we can send at most $\cong 2^{nI(X;Y)}$ distinguishable sequences of length n, (or $\log 2^{nI(X;Y)} = nI(X;Y)$ bits), which is $I(X;Y)$ bits per channel use. Maximizing this quantity over $p(x)$ yields the capacity C.

11.7   The Additive Gaussian Noise Channel and Capacity

An important model that is a building block for many important communications problems is the additive Gaussian noise channel given by

$$Y_i = X_i + Z_i \tag{11.12}$$

where X_i is the transmitted symbol and the noise $Z_i \sim \eta(0, N)$ is statistically independent of X_i. The input symbol X_i is often subject to an average (or peak) energy or power constraint, which fits many physical situations well. For any codeword $(x_1, x_2, \ldots, x_n)$ to be transmitted over the channel, the average power constraint can be expressed as

$$\frac{1}{n} \sum_{i=1}^{n} x_i^2 \leq P . \tag{11.13}$$

This channel model serves as a tool to get to what are called *sphere packing arguments* that should prove useful for agent learning analyses.

We can define the *information* capacity of the additive Gaussian noise channel with average power constraint P as

$$C = \max_{p(x):EX^2 \leq P} I(X; Y). \tag{11.14}$$

and the capacity can be (and will be) shown to have the form

$$C = \frac{1}{2} \log \left(1 + \frac{P}{N} \right) . \tag{11.15}$$

We give a signal space type of justification for this result based on a sphere packing argument.

Geometric Plausibility Argument: Why we are able the construct $(2^{nC}, n)$ codes with low error probability: Consider any codeword of length n. With high probability, the received vector is contained in a sphere of radius $\sqrt{n(N + \epsilon)}$ around the true codeword. If all received vectors in this sphere are assigned to the given codeword, there will be an error when this codeword is sent only if the received vector falls outside this sphere, which is an event with low probability. We choose the decoding spheres of other codewords similarly.

Those readers familiar with signal space, signal constellations, and error probability calculations in digital communications should note that the preceding and following development is very similar.

How many such codewords can we choose? The volume of an n-dimensional sphere is of the form $A_n r^n$ where r is the radius of the sphere. In our case, each of the decoding spheres has radius $\sqrt{nN}$. The received vectors lie in a sphere of radius $\sqrt{n(P + N)}$. Thus, the maximum number of non-intersecting decoding spheres in this volume is no more than

$$\frac{A_n \left(\sqrt{n(P + N)} \right)^n}{A_n \left(\sqrt{nN} \right)^n} = \left(1 + \frac{P}{N} \right)^{\frac{n}{2}}$$

$$= 2^{\frac{n}{2} \log(1 + \frac{P}{N})} \tag{11.16}$$

so the code rate is $\frac{1}{2} \log(1 + \frac{P}{N})$ bits/channel use. This is called a sphere packing argument.

11.8 Parallel Gaussian Channels

The basic expression for the capacity of an additive Gaussian noise channel is often further exploited to obtain very interesting and important results for more interesting channels. Parallel independent Gaussian channels with a common power constraint and channels with colored Gaussian noise play an important role in practical communications and compression problems. The capacity of parallel channels is studied in this section and colored noise channels in the following section. It is envisioned that these channel models and associated results can serve as models for learning systems with multiple parallel paths.

Here we consider independent Gaussian channels in parallel with a common power constraint. The objective is to distribute the total power among the channels so as to maximize the capacity. This models a non-white additive Gaussian noise channel where each parallel component represents a different frequency.

For each parallel channel j, $Y_j = X_j + Z_j$ with $Z_j \sim \eta(0, N_j)$ and the noise is assumed independent from channel to channel. The common power constraint is $E \sum_{i=1}^{k} X_i^2 \le P$. We wish to distribute the power among the various channels so as to maximize the total capacity. The information capacity of the parallel channels is

$$C = \max_{f(x_1,\dots,x):\sum EX_i^2 \le P} I(X_1,\dots,X_k;Y_1,\dots,Y_k). \qquad (11.17)$$

We calculate the distribution that achieves the information capacity for this channel (Note that the term Information Capacity corresponds to the expression for channel capacity based on the maximization of mutual information. This term is used for this quantity before a channel coding theorem is proved, thus giving the information capacity physical meaning).

Since $Z_1, Z_2, \dots, Z_k$ are independent and independent of $X_1, \dots, X_k$,

$$I(X_1,\dots,X_k;Y_1,\dots,Y_k) = h(Y_1,\dots,Y_k) - h(Y_1,\dots,Y_k|X_1,\dots,X_k)$$

$$= h(Y_1,\dots,Y_k) - \sum_{i=1}^{k} h(Z_i)$$

$$\le \sum_{i=1}^{k} \left(h(Y_i) - \sum h(Z_i) \right)$$

$$\le \sum_{i=1}^{k} \frac{1}{2} \log \left(1 + \frac{P_i}{N_i} \right), \qquad (11.18)$$

where $P_i = EX_i^2$ and $\sum\limits_i P_i = P$. Equality is achieved by

$$(X_1,\dots,X_k) \sim \eta \left(0, \begin{bmatrix} P_1 & 0 & \dots & 0 \\ 0 & P_2 & \dots & \dots \\ \vdots & \vdots & \vdots & \vdots \\ 0 & \dots & 0 & P_k \end{bmatrix} \right).$$

The problem is thus reduced to finding the power allotment that maximizes the capacity subject to the constraint $\sum P_i = P$. Using Lagrange multipliers,

$$J(P_1,\dots,P_k) = \sum \frac{1}{2} \log \left(1 + \frac{P_i}{N_i} \right) + \lambda \left(\sum P_i - P \right).$$

Differentiating with respect to P_j,

$$\frac{\partial}{\partial P_j} J(P_1,\dots,P_k) = \frac{1}{2} \frac{1}{\left(1 + \frac{P_j}{N_j} \right)} \cdot \left(\frac{1}{N_j} \right) + \lambda$$

$$= \frac{1}{2} \cdot \frac{1}{Pj + N_j} + \lambda = 0.$$

Solving for P_j, $\frac{1}{P_j + N_j} = -2\lambda$ or $P_j + N_j = \frac{-1}{2\lambda}$ so $P_j = \nu - N_j$.

We must impose the conditions

$$\sum_i P_i = P \text{ and } P_i \geq 0 \tag{11.19}$$

so

$$\sum P_j = \sum (v - N_j) = kv - \sum_{j=1}^{k} N_j = P \tag{11.20}$$

which yields

$$v = \frac{1}{k} \left(P + \sum_{j=1}^{k} N_j \right) = (P + N)_{\text{avg}}.$$

However, the P_i must be non-negative and a solution of this form may not exist. In this case we use the Kuhn–Tucker conditions to verify that the solution that maximizes capacity is

$$P_i = (v - N_i)^+ \tag{11.21}$$

where

$$(x)^+ = \begin{Bmatrix} x, & x \geq 0 \\ 0, & x < 0 \end{Bmatrix}$$

and where v is chosen so that

$$\sum_{i=1}^{k} (v - N_i)^+ = P. \tag{11.22}$$

The Kuhn–Tucker conditions imply that

$$\frac{1}{2} \frac{1}{P_j + N_j} + \lambda = \Lambda \text{ for } P_j > 0$$

or

$$P_j + N_j = \frac{1}{2(\Lambda - \lambda)} \text{ for } P_j > 0.$$

But

$$\frac{1}{2} \frac{1}{P_j + N_j} + \lambda \leq \Lambda \text{ for } P_j = 0$$

or

$$N_j \geq \frac{1}{2(\Lambda - \lambda)}.$$

The implication of the above result is that to maximize capacity, the power should be allocated to less-noisy channels first and then successively to other parallel channels. If a channel is too noisy, then we do not transmit information, that is we do not allocate any power, to this channel at all. This result can be seen to be an illustration of what is called the water filling approach, since transmitted power is allocated to channels until all channels have a noise plus transmitted power amount at the constraint level.

11.9   Channels with Colored Gaussian Noise

We consider parallel Gaussian channels where the noise is dependent. For channels with memory, we can consider a block of n consecutive uses of the channels as n channels in parallel with dependent noise. We develop an expression for capacity, sometimes called the information capacity, but do not prove a coding theorem to establish its physical interpretation as channel capacity. These proofs are left to the references.

Let K_z and K_x be the covariance matrices of the noise and the input, respectively. The input for the constraint is

$$\frac{1}{n} \sum_i E X_i^2 = \frac{1}{n} \operatorname{tr}(K_x) \leq P \tag{11.23}$$

The power constraint depends on n so the capacity must be calculated for each n. We can write

$$I(X_1, X_2, \ldots, X_n; Y_1, \ldots, Y_n) = h(Y_1, Y_2, \ldots, Y_n)$$
$$- h(Z_1, Z_2, \ldots, Z_n). \tag{11.24}$$

$\{Z_i\}$ is not dependent on $\{X_i\}$, so finding capacity amounts to maximizing $h(Y_1, \ldots, Y_n)$. This is maximum for Gaussian Y which is achieved when X is Gaussian. Thus, since the input and noise are independent, $K_y = K_x + K_z$ and

$$h(Y_1, \ldots, Y_n) = \frac{1}{2} \log \left[(2\pi e)^n |K_x + K_z| \right] . \tag{11.25}$$

The problem is thus reduced to choosing K_x to maximize $|K_x + K_z|$, subject to the power constraint $\operatorname{tr}(K_x)$ constraint. Write

$$K_z = Q \Lambda Q^T \quad \text{when} \quad Q Q^T = I .$$

Then

$$|K_x + K_z| = \left| K_x + Q \Lambda Q^T \right|$$
$$= |Q| \left| Q^T K_x Q + \Lambda \right| \left| Q^T \right|$$
$$= \left| Q^T K_x Q + \Lambda \right| = |A + \Lambda| \tag{11.26}$$

where $A = Q^T K_x Q$. For any matrices C and B, $\operatorname{tr}(BC) = \operatorname{tr}(CB)$, so

$$\text{tr}\,(A) = \text{tr}\,Q^T K_x Q = \text{tr}\,\left(QQ^T K_x\right)$$
$$= \text{tr}\,K_x \tag{11.27}$$

and we can maximize $|A + \Lambda|$ subject to trace constraint on K_x.

By Hadamard's inequality

$$|K| \leq \Pi_i K_{ii}$$

with equality iff K is diagonal, and also we have

$$|A + \Lambda| \leq \Pi_i (A_{ii} + \lambda_i) \tag{11.28}$$

with equality iff A is diagonal. A is also subject to the constraints $\frac{1}{n} \sum_i A_{ii} \leq P$ and $A_{ii} \geq 0$.

Using the Kuhn–Tucker conditions on

$$J = h(Y_1, \ldots, Y_n) + \gamma\left(\frac{1}{n}\sum A_{ii} - P\right)$$

so

$$\frac{1}{[A_{jj} + \lambda_j]} = 2\left[\Gamma - \frac{\gamma}{n}\right]$$

or

$$[A_{jj} + \lambda_j] = \frac{1}{2\left[\Gamma - \frac{\gamma}{n}\right]} = \nu$$

for $A_{jj} > 0$, and

$$\frac{1}{\lambda_j} \leq 2\left[\Gamma - \frac{\gamma}{n}\right]$$

or

$$\lambda_j \geq \frac{1}{2\left[\Gamma - \frac{\gamma}{n}\right]} = \nu \text{ for } A_{jj} = 0.$$

This says pick $A_{jj} = 0$ if $\lambda_j > \nu$, otherwise choose A_{jj} to satisfy $A_{jj} + \lambda_j = \nu$. These conditions can be expressed as

$$A_{jj} = [\nu - \lambda_j]^+.$$

Choose ν to satisfy $\frac{1}{n}\sum_i A_{ii} = P$ or

$$\frac{1}{n}\sum_i A_{ii} = \frac{1}{n}\left[n\nu - \sum \lambda_i\right] = P$$

or

$$P = \nu - \frac{1}{n}\sum \lambda_i.$$

The simple independent parallel Gaussian channel result has now been extended to the even more interesting case of parallel Gaussian channels where the noise is dependent. In many applications when we do not know the distribution of the channel noise, we can still use this result to obtain substantive results.

Rate Distortion Theory — 12

12.1 Introduction

The rate distortion function is the minimum rate that we can represent the (source) sequence with average distortion D. The collection of topics in this chapter contains many classical results which have found utility or mention in agent learning investigations. However, the coding theorem oriented proofs are not provided, since these are in standard information theory texts and also because research in that direction is not appropriate for agent learning performance analyses.

A number of topics in this chapter are not usually covered in a first course in information theory. Among these topics are a treatment of the Shannon Lower Bound and the optimum backward channel, a development of successive refinement of information, and an introduction to composite source models, which are included because of their existence in some information theoretic learning papers and their possible utility in future research.

12.2 Definition of the Rate Distortion Function

For a discrete source, or any set of sequences, subject to an average distortion constraint, the rate distortion function requires the solution of the following minimization problem by selecting the conditional probabilities $p(\hat{x}|x)$

$$R(D) = \min_{p(\hat{x}|x):\sum_{(x,\hat{x})} p(x)p(\hat{x}|x)d(x,\hat{x})\leq D} I(X; \hat{X}). \tag{12.1}$$

The conditional probabilities $p(\hat{x}|x)$ describe the probabilistic mapping of the source to the corresponding reconstructed output.

© The Author(s), under exclusive license to Springer Nature Switzerland AG 2025   77
J. D. Gibson, *Information Theoretic Principles for Agent Learning*, Synthesis Lectures on Engineering, Science, and Technology, https://doi.org/10.1007/978-3-031-65388-9_12

Expanding the mutual information, we have

$$I\left(X;\hat{X}\right) = \sum_x \sum_{\hat{x}} p(x) p(\hat{x}|x) \log \frac{p(\hat{x}|x)}{p(\hat{x})}. \tag{12.2}$$

which is to be minimized subject to the constraint on average distortion

$$\bar{d}(x,\hat{x}) = \sum_x \sum_{\hat{x}} p(x) p\left(\hat{x}|x\right) d\left(x,\hat{x}\right), \tag{12.3}$$

To state the minimization succinctly we define the set

$$\mathcal{P}_D = \left\{ p\left(\hat{x}|x\right) : \bar{d}(x,\hat{x}) \leq D \right\}. \tag{12.4}$$

Thus, we have

$$R(D) = \min_{p(\hat{x}) \in \mathcal{P}_D} I\left(X;\hat{X}\right) \tag{12.5}$$

12.3   Solving the Constrained Minimization Problem

To perform the minimization, we use Lagrange multipliers to append the constraints, so the functional to be minimized becomes

$$J(p(\hat{x}|x)) = \sum_x \sum_{\hat{x}} p\left(x\right) p\left(\hat{x}|x\right) \log \frac{p\left(\hat{x}|x\right)}{p\left(\hat{x}\right)}$$
$$+ \lambda \sum_x \sum_{\hat{x}} p(x) p\left(\hat{x}|x\right) d\left(x,\hat{x}\right) \tag{12.6}$$
$$+ \sum_x \beta(x) \sum_{\hat{x}} p(\hat{x}|x)$$

The minimization can be performed by taking partial derivatives with respect to the conditional density $p(\hat{x}|x)$ or by variational calculus to obtain the form for the conditional density

$$p(\hat{x}|x) = \frac{p(\hat{x}) \exp\left[-\lambda d(x,\hat{x})\right]}{\Gamma(x)} \tag{12.7}$$

where

$$\Gamma(x) = \sum_{\hat{x}} p(\hat{x}) \exp\left[-\lambda d(x,\hat{x})\right] \tag{12.8}$$

the latter of which imposes the property that the conditional pdf sums to 1.

Multiplying Eq. (12.7) by $p(x)$ and summing over all x and simplifying by assuming $p(\hat{x}) > 0$, we find

$$\sum_x \frac{p(x) \exp\left[-\lambda d(x,\hat{x})\right]}{\sum_{\hat{x}'} p(\hat{x}') \exp\left[-\lambda d(x,\hat{x}')\right]} = 1 \tag{12.9}$$

Equation (12.9) holds for $\hat{x} \in \hat{\chi}$, which gives $|\hat{\chi}|$ equations which when combined with the average distortion constraint allows us to evaluate the Lagrange multiplier λ and $p(\hat{x})$. We then use Eq. (12.7) and the Kuhn-Tucker conditions to obtain the conditions for the minimum.

12.4   The Rate Distortion Function for Continuous Amplitude Sources

For a discrete-time continuous amplitude source with single-letter distortion measure $d(u, z)$, each conditional pdf relating the source output to the user input produces an average distortion given by [2, 3]

$$\bar{d}\left(f_{Z|U}\right) = \int_{-\infty}^{\infty} \int_{-\infty}^{\infty} f_U(u) f_{Z|U}(z|u) d(u, z) du\, dz \qquad (12.10)$$

and a mutual information

$$I(U; Z) = \int_{-\infty}^{\infty} \int_{-\infty}^{\infty} f_U(u) f_{Z|U}(z|u) \log \frac{f_{Z|U}(z|u)}{f_Z(z)} du\, dz . \qquad (12.11)$$

The admissible pdfs are described by the set

$$\mathcal{P}_D = \left\{ f_{Z|U}(z|u) : \bar{d}\left(f_{Z|U}\right) \leq D \right\} .$$

The rate distortion function is then defined as[1]

$$R(D) = \min_{f_{Z|U} \in \mathcal{P}_D} I(U; Z) . \qquad (12.12)$$

A significant difference between the rate distortion functions for discrete-amplitude and continuous-amplitude sources is that for $R(D)$ in Eq. (12.12), as $D \to 0$, $R(D) \to \infty$.

Analytical calculation of the rate distortion function for continuous-amplitude sources often is extremely difficult, and relatively few such calculations have been accomplished. We present the results of one such calculation in the following example. For the squared-error distortion measure

$$d(u - z) = (u - z)^2 , \qquad (12.13)$$

a discrete-time, memoryless Gaussian source with zero mean and variance σ^2 has the rate distortion function

$$R(D) = \begin{cases} \frac{1}{2} \log \frac{\sigma^2}{D}, & 0 \leq D \leq \sigma^2 , \\ 0, & D > \sigma^2 . \end{cases} \qquad (12.14)$$

---

[1] Strictly speaking, the "min" in Eq. (12.12) should be replaced with "inf," denoting infimum or greatest lower bound.

Let X be $\sim N(0, \sigma^2)$. By the rate distortion theorem, we have

$$R(D) = \min_{f(\hat{x}|x):E(\hat{X}-X)^2 \leq D} I(X; \hat{X}) . \tag{12.15}$$

We first find a lower bound for the rate distortion function and then guess a distribution to show that the lower bound is achievable. We know that $E(X - \hat{X})^2 \leq D$, so we have the following

$$
\begin{aligned}
I(X; \hat{X}) &= h(X) - h(X|\hat{X}) \\
&= \frac{1}{2} \log(2\pi e)\sigma^2 - h(X - \hat{X}|\hat{X}) \\
&\geq \frac{1}{2} \log(2\pi e)\sigma^2 - h(X - \hat{X}) \tag{12.16}
\end{aligned}
$$

since conditioning cannot increase entropy, and continuing

$$\geq \frac{1}{2} \log(2\pi e)\sigma^2 - h(N(0, E(X - \hat{X})^2)) \tag{12.17}$$

because the Gaussian distribution maximizes entropy,

$$
\begin{aligned}
&= \frac{1}{2} \log(2\pi e)\sigma^2 - \frac{1}{2} \log(2\pi e) E(X - \hat{X})^2 \\
&\geq \frac{1}{2} \log(2\pi e)\sigma^2 - \frac{1}{2} \log(2\pi e) D \tag{12.18}
\end{aligned}
$$

invoking the condition that $E(X - \hat{X})^2 \leq D$,

$$= \frac{1}{2} \log \frac{\sigma^2}{D} , \tag{12.19}$$

where (12.4) follows from the fact that conditioning reduces entropy and (12.17) follows from the fact that the normal distribution maximizes the entropy for a given second moment. Hence

$$R(D) \geq \frac{1}{2} \log \frac{\sigma^2}{D} . \tag{12.20}$$

Therefore, the rate distortion function for the Gaussian source with squared error distortion is

$$R(D) = \begin{cases} \frac{1}{2} \log \frac{\sigma^2}{D}, & 0 \leq D \leq \sigma^2 , \\ 0, & D > \sigma^2 . \end{cases} \tag{12.21}$$

as illustrated in Fig. 12.1. Equation (12.21) is a relatively simple result that is given added importance by the fact that the Gaussian source is a worst-case source in the sense that it requires the maximum rate of all possible sources to achieve a specified mean square-

Fig. 12.1 R(D) for memoryless Gaussian source and squared error distortion measure

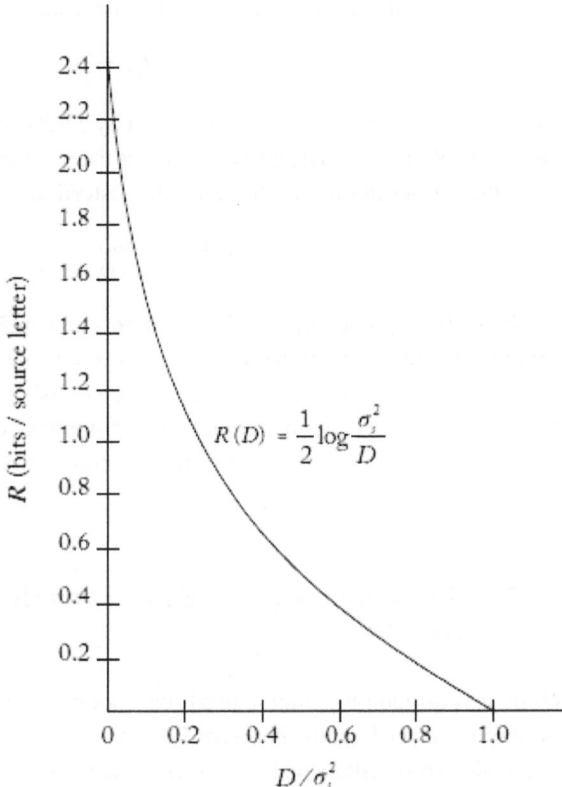

error distortion. Specifically, any memoryless, zero-mean, continuous-amplitude source with variance σ_s^2 has a rate distortion function $R(D)$ with respect to the squared-error distortion measure that is upper bounded as

$$R(D) \leq \tfrac{1}{2} \log \frac{\sigma_s^2}{D}, \qquad 0 \leq D \leq \sigma_s^2 . \tag{12.22}$$

This result implies that if the Gaussian source with the chosen average distortion can be transmitted over the channel, then any other memoryless source with the same variance will be supported by the same channel with the same average distortion. However, it may be possible to transmit at a lower rate for the given source if it is non-Gaussian. Of course, to find the required rate, the true source distribution must be known and an expression for its rate distortion function must be found. These conditions are not always easy to satisfy, so the Gaussian result is very useful.

We can rewrite (12.21) to express the distortion in terms of the rate,

$$D(R) = \sigma^2 2^{-2R} \ . \tag{12.23}$$

The approach to compression represented by $D(R)$ should be contrasted with the approach used to develop $R(D)$. The latter minimizes the rate for a given (chosen) distortion constraint, while the former implies minimizing the distortion subject to a rate constraint, that is,

$$D(R) = \min_{f(\hat{x}|x): I(X;\hat{X}) \leq R} E(\hat{X} - X)^2 \ . \tag{12.24}$$

The $R(D)$ approach fits problems where a known distortion is acceptable and we need to minimize the rate. This approach fits audio coding and video coding where near-transparent quality is desired and the rate needed to accomplish this goal is accepted by the application. The $D(R)$ approach fits many network or wireless communications problems where the maximum rate available is fixed and the compression is accomplished to minimize the distortion.

12.5   The Shannon Lower Bound and the Optimum Backward Channel

While it is difficult to obtain closed form solutions for $R(D)$ in general, there have been a number of lower bounds that have been developed. The most useful for us and for many applications is the Shannon lower bound (SLB) for difference distortion measures derived by Shannon in his 1959 paper [44]. For the discrete case, Shannon provided the following proof.

For a source X and its reconstruction $\hat{X}$ that achieves an average distortion $Ed(X, \hat{X}) \leq D$ for some difference distortion measure, we expand $I(X; \hat{X})$ as

$$
\begin{aligned}
I(X; \hat{X}) &= H(X) - H(X|\hat{X}) \\
&= H(X) - \sum_{\hat{x}} p(\hat{x}) H(X|\hat{X} = \hat{x}) \\
&\geq H(X) - \sum_{\hat{x}} p(\hat{x}) \phi(D_{\hat{x}})
\end{aligned}
\tag{12.25}
$$

where we have used the notation $\phi(D_{\hat{x}})$ to denote the maximum of the conditional entropy $H(X|\hat{X} = \hat{x})$ that achieves $D_{\hat{x}} = \sum_x p(x|\hat{x})d(x, \hat{x})$. We next use Jensen's inequality to obtain

$$
\begin{aligned}
I(X; \hat{X}) &\geq H(X) - \phi \left(\sum_{\hat{x}} p(\hat{x}) D_{\hat{x}} \right) \\
&\geq H(X) - \phi(D) \ ,
\end{aligned}
\tag{12.26}
$$

where the last expression follows since $D_{\hat{x}} \leq D$.

These are basically the set of steps we followed in developing the lower bounds on $R(D)$ for the memoryless Gaussian source in the preceding section.

To find the condition where the lower bound produced here and the bounds obtained for these specific sources is satisfied with equality, Shannon showed that for a continuous source X and a difference distortion measure $d(x, \hat{x}) = d(x - \hat{x})$ subject to the constraint

$$\iint p(x)p(\hat{x}|x)d(x - \hat{x}) \, dx \, d\hat{x} \leq D . \tag{12.27}$$

then

$$R(D) \geq h(X) + \int g(e) \log g(e) \, de \tag{12.28}$$

or

$$R(D) \geq h(X) - \max h(g(e)), \tag{12.29}$$

where $g(e) = g(x - \hat{x})$ and the maximum is taken over all probability densities for the error that satisfy the average distortion constraint [3]. Equality is achieved iff

$$\int p(\hat{x})g(x - \hat{x}) \, d\hat{x} = p(x). \tag{12.30}$$

The equality condition produces the Shannon optimum backward channel, which states that the reconstructed value and the encoding error are statistically independent and sum to the input source value. It is the statistical independence result via the Shannon optimum backward channel that allows us to obtain the final form of $R(D)$ for the memoryless Gaussian source given earlier.

12.6 Reverse Water-Filling

From the equality condition in the Shannon lower bound for difference distortion measures, we know that the reconstructed value, the source, and the encoding error must satisfy the optimum Shannon backward channel result, $X = \hat{X} + Z$, where $\hat{X}$ and Z are statistically independent. Since the source X is Gaussian with mean zero and variance σ^2 and the variance of the error is D, and the probability density for the error that maximizes the entropy is Gaussian, we find that $Z \sim N(0, D)$ and $\hat{X} \sim N(0, \sigma^2 - D)$.

We have directly for these distributions that

$$I(X; \hat{X}) = \frac{1}{2} \log \frac{\sigma^2}{D} , \tag{12.31}$$

for $E(X - \hat{X})^2 = D$ and $D \leq \sigma^2$, and $R(D) = 0$ if $D > \sigma^2$, thus achieving the bound in (12.20).

The rate distortion function of a vector X of independent (but not identically distributed) Gaussian sources is calculated by the reverse water-filling theorem. This theorem says that we should encode the independent subsources X_i with equal distortion level λ, as long as λ does not exceed the variance of the transmitted subsource, and that one should not transmit at all those subsources whose variance is less than the distortion λ.

(Reverse water-filling theorem) For a vector X of independent random variables $X_1, X_2, ..., X_n$ such that $X_i \sim N(0, \sigma_i^2)$ and the distortion measure

$$D(\underline{X}, \underline{\hat{X}}) = E\left[\sum_{i=1}^{n}(X_i - \hat{X}_i)^2 = \sum_{i=1}^{n} D_i\right] \leq D, \tag{12.32}$$

the rate distortion function is

$$R(D) = \min_{p(\underline{\hat{x}}|x):D(\underline{X},\underline{\hat{X}})\leq D} I(X; \hat{X}) = \sum_{i=1}^{n} \frac{1}{2} \log \frac{\sigma_i^2}{D_i}, \tag{12.33}$$

where

$$D_i = \begin{cases} \lambda & 0 \leq \lambda \leq \sigma_i^2 \\ \sigma_i^2 & \lambda > \sigma_i^2 \end{cases}. \tag{12.34}$$

To begin, we use standard arguments to simplify the mutual information of the n-vectors as

$$I(X; \hat{X}) = h(X) - h(X|\hat{X}) \tag{12.35}$$

$$= \sum_{i=1}^{n} h(X_i) - \sum_{i=1}^{n} h(X_i|X^{i-1}, \hat{X}) \tag{12.36}$$

$$\geq \sum_{i=1}^{n} h(X_i) - \sum_{i=1}^{n} h(X_i|\hat{X}_i) \tag{12.37}$$

$$= \sum_{i=1}^{n} I(X_i; \hat{X}_i) \tag{12.38}$$

$$\geq \sum_{i=1}^{n} R(D_i) \tag{12.39}$$

$$= \sum_{i=1}^{n} \left(\frac{1}{2} \log \frac{\sigma_i^2}{D_i}\right)^+, \tag{12.40}$$

where the superscript $+$ indicates that the term is zero if the quantity is negative.

From the Shannon backward channel result, we can achieve equality in the lower bound by choosing $f(x^n|\hat{x}^n) = \prod_{i=1}^{n} f(x_i|\hat{x}_i)$ and setting the distribution of each $\hat{X}_i \sim N(0, \sigma_i^2 - D_i)$.

To find the rate distortion function for the n-vector X, we now have the optimization problem

$$R(D) = \min_{\sum D_i = D} \sum_{i=1}^{n} \max \left\{ \frac{1}{2} \ln \frac{\sigma_i^2}{D_i}, 0 \right\} .$$ (12.41)

To apply the Kuhn-Tucker conditions we use Lagrange multipliers and form the functional

$$J(D) = \sum_{i=1}^{m} \frac{1}{2} \ln \frac{\sigma_i^2}{D_i} + \lambda \sum_{i=1}^{m} D_i ,$$ (12.42)

which upon differentiating with respect to D_i, we get

$$\frac{\partial J}{\partial D_i} = -\frac{1}{2} \frac{1}{D_i} + \lambda ,$$ (12.43)

where λ is chosen so that

$$\frac{\partial J}{\partial D_i} \begin{cases} = 0, \text{ if } D_i < \sigma_i^2 , \\ \leq 0, \text{ if } D_i \geq \sigma_i^2 . \end{cases}$$ (12.44)

and $\sum_{i=1}^{m} D_i = D$.

Once we choose a constant λ, we code those subsources with variances greater than λ and spend zero bits on those subsources with a variance less than λ. This gives us the desired $R(D)$ for the n-vector X, which has the reverse water-filling interpretation.

Reverse water-filling is a classical result in rate distortion theory and it plays a major role in many applications.

12.7 Successive Refinement of Information

In applications, it is sometimes useful to first access a low resolution version of a source, and then, if that is the source of interest, to send additional bits in order to improve the resolution. This has been called SNR scalability, successive approximation, or progressive encoding. In terms of rate distortion theory, this approach as has been called Successive Refinement. A source X with distortion measure $d_1(X, \hat{X})$ is said to be successively refinable if and only if the successive approximation falls exactly on the rate distortion function $R(D)$ for this source and distortion measure and single full rate transmissions.

More specifically, consider a sequence of random variables $X_1, \ldots, X_n$ that is successively refined by a two-stage description that is rate distortion optimal at each stage. The X sequence is encoded as $\hat{X}$ at rate R_1 bits per symbol with average distortion D_1. Then information is added to the first message at rate $R_e = R_2 - R_1$ bits per symbol so that the resulting two-stage reconstruction $\hat{X}_r$ now has average distortion D_2 at rate $R_2 \geq R_1$.

A generalization of both successive refinement and successive approximation in terms of rate distortion theory is the concept of a rate distortion region. The following theorem specifies the rate distortion region for a discrete memoryless source X and two distortion measures d_1 and d_2.

Theorem: The successive refinement rate distortion region $R(D_1, D_2)$ for a DMS X and distortion measures d_1 and d_2 is the set of rate pairs (R_1, R_2) such that

$$R_1 \geq I(X; \widehat{X}_1)$$
$$R_1 + R_2 \geq I(X; \widehat{X}_1, \widehat{X}_2) \tag{12.45}$$

for some conditional probability mass function $p(\hat{x}_1, \hat{x}_2|x)$ that satisfies the constraints $E[d_1(X, \widehat{X}_1)] \leq D_1$ and $E[d_2(X, \widehat{X}_2)] \leq D_2$.

The achievability of this rate distortion region is proved by Rimoldi, who also uses classical arguments to present a clear explanation about the difference between successive approximation and successive refinement. That is, given a source X, we only are concerned with the approximately $2^{nH(X)}$ corresponding typical sequences. We can partition this typical set into balls of radius D_1 with cardinality $2^{nH(X|\widehat{X}_1)}$ if there exists a random variable $\hat{X}_1$ such that $E[d_1(X, \widehat{X}_1)] \leq D_1 \cdot \frac{2^{nH(X)}}{2^{nH(X|\hat{X}_1)}}$ of these balls are required to cover the set of typical sequences.

Letting $\hat{x}_i$ be an output in the ith ball, then the next step is to cover the ith ball with balls of smaller radius D_2. As before, we can construct balls of radius D_2 with cardinality of approximately $2^{nH(X|\hat{X}_2)}$ if there exists a random variable $\hat{X}_2$ such that $E[d_2(X, \widehat{X}_2)] \leq D_2$. It would take approximately $\frac{2^{nH(X|\hat{X}_1)}}{2^{nH(X|\hat{X}_2)}}$ to cover the large balls of radius D_1 if all of the smaller diameter balls fit exactly into the larger balls.

However, in general, a small ball may only intersect. a larger ball, and thus the intersection has cardinality of approximately $2^{nH(X|\widehat{X}_1, \widehat{X}_2)}$. Since $2^{nH(X|\widehat{X}_1, \widehat{X}_2)} \leq 2^{nH(X|\hat{X}_2)}$, the rate is higher since it will take

$$2^{[nH(X|\hat{X}_1) - nH(X|\hat{X}_1, \hat{X}_2)]} = 2^{nI(X; \widehat{X}_2|\hat{X}_1)} \tag{12.46}$$

small balls to cover the larger balls. In this case, we have successive approximation but we do not fall on the rate distortion function.

We see that we will only achieve the lower rate if

$$H(X|\widehat{X}_1, \widehat{X}_2) = H(X|\hat{X}_2) \tag{12.47}$$

or when

$$X \to \widehat{X}_1 \to \widehat{X}_2 \tag{12.48}$$

forms a Markov chain. This is the case that falls on the rate distortion function and hence is designated as successively refinable. When Eq. (12.47) holds then

$$H(X|\widehat{X}_1) - H(X|\widehat{X}_2) = I(X; \widehat{X}_2) - I(X; \widehat{X}_1)$$
$$= I(X; \widehat{X}_2|\widehat{X}_1) \tag{12.49}$$

which by the chain rule of mutual information verifies that Eq. (12.45) is satisfied with equality and thus we have successive refinement.

Theorem: [23, 24] Successive refinement with distortion D_1 and D_2 ($D_1 \geq D_2$) can be achieved if and only if there exists a conditional distribution $p(\hat{x}, \hat{x}_r \mid x)$ with $Ed(X, \hat{X}) \leq D_1$ and $Ed(X, \hat{X}_r) \leq D_2$, such that $R(D_1) = I(X; \hat{X})$ and $R(D_2) = I(X; \hat{X}_r)$ and $p(\hat{x}, \hat{x}_r \mid x) = p(\hat{x}_r \mid x)p(\hat{x} \mid \hat{x}_r)$.

◇

The last condition is equivalent to saying that X, $\hat{X}$, $\hat{X}_r$ can be written as the Markov chain $X \to \hat{X}_r \to \hat{X}$, or, equivalently, as $\hat{X} \to \hat{X}_r \to X$. According to the generalization by Rimoldi, we can extend this result for different distortion measures at each layer as follows [23].

Theorem: Successive refinement with distortion D_1 and D_2 ($D_1 \geq D_2$) can be achieved if and only if there exists a conditional distribution $p(\hat{x}, \hat{x}_r \mid x)$ with $Ed_1(X, \hat{X}) \leq D_1$ and $Ed_2(X, \hat{X}_r) \leq D_2$, such that $R_1 \geq I(X; \hat{X})$ and $R_2 \geq I(X; \hat{X}, \hat{X}_r)$.

◇

Lemma: Successive refinement with distortion D_1 and D_2 ($D_1 \geq D_2$) can be achieved if there exists a conditional distribution $p(\hat{x}, \hat{x}_r \mid x)$ with $Ed_1(X, \hat{X}) \leq D_1$ and $Ed_2(X, \hat{X}_r) \leq D_2$, such that $R_1 \geq I(X; \hat{X})$ and $R_2 - R_1 \geq I(X; \hat{X}_r \mid \hat{X})$.

See Figs. 12.2 and 12.3.

Fig. 12.2 Two stage successive refinement

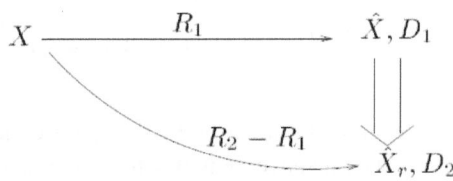

Fig. 12.3 Successive refinement of information

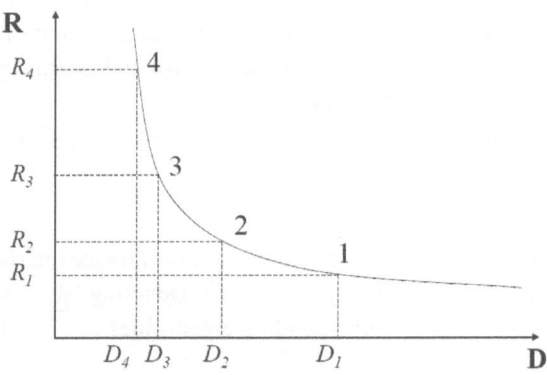

12.8   Stationary Gaussian Sources with Memory

In this section, we present a classic derivation recasting the rate distortion problem for stationary Gaussian sources into a form that is easy to use in many applications. This result apparently first appeared in [10].

Let A be an unitary matrix denoting an orthonormal linear transformation from a vector of random variables $\underline{X}$ to another vector of random variables $\underline{\Theta}$ as

$$\underline{\Theta} = A\underline{X}, \; \hat{\underline{X}} = A^{-1}\hat{\underline{\Theta}} = A^T \hat{\underline{\Theta}}. \tag{12.50}$$

The following relations between $\underline{X}$ and $\underline{\Theta}$ can be derived:

Mean squared error:

$$
\begin{aligned}
D(\underline{X}, \hat{\underline{X}}) \quad &= \quad E[(\underline{X} - \hat{\underline{X}})^T (\underline{X} - \hat{\underline{X}})] \\
&= \quad E[(\underline{\Theta} - \hat{\underline{\Theta}})^T A^T A (\underline{\Theta} - \hat{\underline{\Theta}})] \\
&= \quad E[(\underline{\Theta} - \hat{\underline{\Theta}})^T (\underline{\Theta} - \hat{\underline{\Theta}})] \\
\text{orthonormal} \quad &= \quad D(\underline{\Theta}, \hat{\underline{\Theta}});
\end{aligned}
\tag{12.51}
$$

Mutual information:

$$I(\underline{X}; \hat{\underline{X}}) \underset{|A| \neq 0}{=} I(\underline{\Theta}; \hat{\underline{\Theta}}) \geq \sum_{i=1}^{n} I(\Theta_i; \hat{\Theta}_i), \tag{12.52}$$

with equality if and only if (iff) Θ_i's are independent.

As shown above, both the distortion (chosen here to be the summation of squared errors) and the mutual information of a random process are equal to those of the unitary transform of the random process $\underline{\Theta}$, and therefore the rate distortion function of $\underline{X}$ equals the rate distortion function of $\underline{\Theta}$.

Therefore, to utilize the reverse-water-filling result, we need to find $\underline{X}$'s unitary transform $\underline{\Theta}$ with independent elements. This leads us quite naturally to the well known Karhunen Lòeve Transform (KLT), which is also called principal component analysis, to decorrelate the source $\underline{X}$ as follows.

With the covariance function of a stationary Gaussian source denoted by

$$\phi(n) = E[x_i x_{i+n}], \tag{12.53}$$

and with Φ_n representing the $n \times n$ covariance matrix of the source, with entries $\phi(n)$, then, $\Phi_n = \{\phi(|i - j|), i, j = 1, ..., n\}$. Denoting $\{\underline{\psi}_i, i = 1, ..., n\}$ as the normalized eigenvectors of Φ_n with corresponding eigenvalues $\{\lambda_i, i = 1, ..., n\}$, then

$$\Phi_n \underline{\psi}_i = \lambda_i \underline{\psi}_i, \tag{12.54}$$

and

$$\Phi_n = \Psi_n \Lambda \Psi_n^T, \tag{12.55}$$

where $\Psi_n = [\underline{\psi}_1, \underline{\psi}_2, ..., \underline{\psi}_n]$.

Since covariance matrices are symmetric, there always exists an eigenvalue decomposition of the covariance matrix with real eigenvalues, and furthermore, covariance matrices are positive semi-definite, therefore all their eigenvalues are non-negative, yielding

$$\underline{\Theta} = \Psi_n^T \underline{X}. \tag{12.56}$$

Thus, the rate distortion function of a stationary Gaussian source with covariance matrix Φ_n can be computed as the rate distortion function of a stationary Gaussian source $\underline{\Theta}$, where $\underline{\Theta}$ has independent Gaussian elements, each of variance λ_i, which are eigenvalues of the covariance matrix Φ_n. The rate distortion function of $\underline{\Theta}$ is in turn solved by the reverse-water filling theorem.

12.9   Composite Source Models and Conditional Rate Distortion Functions

The idea that sources may have multiple modes and can switch between modes probabilistically, was initially presented by Berger in his classic book, wherein he denoted such sources as composite sources [3]. For a composite source, the choice of subsources is accomplished according to a probabilistic switch process, which is the side information Y [3, Sect. 6.1]. The power of such composite source models is that the individual subsources are able to capture local or fine dependence, while the switch process can represent changes that happen more globally as well as capturing model discontinuities. Given an appropriate number of carefully selected subsources and accurate switch modeling, time-varying or spatially-varying complex real world sources, such as voice and video, can be represented accurately.

A composite source with switch probability depending on the side information Y [3] is represented in Fig. 12.4.

Motivated by the work of Berger [3] and others, research efforts explored, from the theoretical side, the properties of composite sources and also attempted to obtain expressions or bounds for the rate distortion functions of composite sources. The conditional rate distortion function is the rate of a source subject to a fidelity criterion when the encoder and decoder both have access to side information [9]. Thus, the conditional rate distortion function describes the rate required for a composite source subject to a fidelity criterion, where the side information is the switch process that selects the appropriate subsource at any time. The following definition of the conditional rate distortion function is from Gray [9].

The conditional rate distortion function of a source $\underline{X}$ with side information Y, which serves as the subsource information, is defined as

$$R_{\underline{X}|Y}(D) = \min_{p(\hat{\underline{x}}|\underline{x}, y): D(\underline{X}, \hat{\underline{X}}|Y) \leq D} I(\underline{X}; \hat{\underline{X}}|Y), \tag{12.57}$$

Fig. 12.4 A composite source model with K subsources

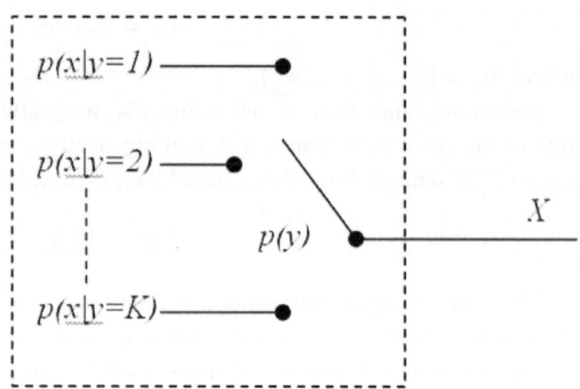

where

$$D(\underline{X}, \hat{\underline{X}}|Y) = \sum_{\underline{x},\hat{\underline{x}},y} p(\underline{x}, \hat{\underline{x}}, y) D(\underline{x}, \hat{\underline{x}}|y), \tag{12.58}$$

and

$$I(\underline{X}; \hat{\underline{X}}|Y) = \sum_{\underline{x},\hat{\underline{x}},y} p(\underline{x}, \hat{\underline{x}}, y) \log \frac{p(\underline{x}, \hat{\underline{x}}|y)}{p(\underline{x}|y) p(\hat{\underline{x}}|y)}. \tag{12.59}$$

$R_{\underline{X}|Y}(D)$ is the lowest rate given that both encoder and decoder are allowed undistorted access to the switch process Y. It can be shown [9] that the conditional rate distortion function in Eq. (12.57) can also be expressed as

$$R_{\underline{X}|Y}(D) = \min_{D'_y s : D(\underline{X}, \hat{\underline{X}}|Y) = \sum_y D_y p(y) \leq D} \sum_y R_{\underline{X}|y}(D_y) p(y), \tag{12.60}$$

and the minimum is achieved by adding up the individual, also called marginal, rate-distortion functions at points of equal slopes of the marginal rate distortion functions. Utilizing the prior expression for conditional rate distortion functions, the minimum is achieved at D_y's where the slopes $\frac{\partial R_{\underline{X}|Y=y}(D_y)}{\partial D_y}$ are equal for all y and $\sum_y D_y P[Y = y] = D$.

This conditional rate distortion function $R_{\underline{X}|Y}(D)$ can be used to write the following inequality involving the overall source rate distortion function $R_{\underline{X}}(D)$ [9]

$$R_{\underline{X}|Y}(D) \leq R_{\underline{X}}(D) \leq R_{\underline{X}|Y}(D) + I(\underline{X}; Y), \tag{12.61}$$

where $I(\underline{X}; Y)$ is the mutual information between $\underline{X}$ and Y and the equality in the leftmost inequality is achieved if and only if $\underline{X}$ and Y are independent. We can simplify the above and further bound $I(\underline{X}; Y)$ by

$$I(\underline{X}; Y) \leq H(Y) \leq \frac{1}{M} \log K, \tag{12.62}$$

where K is the number of subsources and M is the number of samples representing how often the subsources change. For many applications, such as for voice and video coding, K is much smaller than M so that the second term on the right involving the rate of sending the side information is negligible, and the rate distortion for the source is very close to the conditional rate distortion function.

12.10 The Rate Distortion Theorem for Independent Gaussian Sources–Revisited

A mathematical characterization of the rate distortion function for Gaussian sources is given by the following theorem [2, 3].

(Shannon's third theorem–Gaussian Sources) For a vector X of independent random variables X_1, X_2, ..., X_n and the distortion measure $D(\underline{X}, \hat{\underline{X}}) = E\left[\sum_{i=1}^{n}(X_i - \hat{X}_i)^2 = \sum_{i=1}^{n} D_i\right] \leq D$, the rate distortion function is

$$R(D) = \min_{p(\hat{x}|x):D(\underline{X},\hat{\underline{X}})\leq D} I(X; \hat{X}) = \sum_{i=1}^{n} \frac{1}{2} \log \frac{\sigma_i^2}{D_i}, \qquad (12.63)$$

where $I(X; \hat{X})$ is the mutual information between X and $\hat{X}$. Given the source distribution $p(x)$, the minimization is over all admissible test channels, that is, all $p(\hat{x}|x)$ satisfying the average distortion constraint.

The proof of achievability is long and fairly complicated. It is very similar to the proof of channel capacity given in Chap. 4 except that there is a condition added to joint typicality to include sequences that satisfy the distortion constraint, sometimes called distortion typical sequences. We only set up some notation here and leave the details of the proof to the references.

The proof, using a random coding argument and distortion typicality for encoding and decoding, involves showing there exists an encoder/decoder pair generated according to a test channel $p(\hat{x}|x)$ that achieves the rate distortion pair $(R(D), D)$. Notationally, the encoder and decoder are the functions f_n and g_n, respectively, and we get a length-n code with a codebook of 2^{nR} sequences. Instead of the achievability proof, we use a geometric sphere packing argument to motivate the achievability result.

We have a Gaussian source of variance σ^2 and use the encoder/decoder pair specified to get a $(2^{nR}, n)$ rate distortion code for this source with distortion D, which is a set of 2^{nR} codewords in $\mathbb{R}^n$ such that most source sequences of length n (all those that lie within a sphere of radius $\sqrt{n\sigma^2}$) are within a distance $\sqrt{nD}$ of some codeword. By the sphere packing argument, it is clear that the minimum number of codewords required is

$$2^{nR(D)} = \left(\frac{\sigma^2}{D}\right)^{n/2} . \qquad (12.64)$$

The rate distortion theorem shows that this minimum rate is asymptotically achievable, i.e., that there exists a collection of spheres of radius $\sqrt{nD}$ that cover the space except for a set of arbitrarily small probability.

The proof of the converse to the Rate Distortion Theorem requires showing that the rate R of any rate distortion code exceeds the rate distortion function $R(D)$ evaluated at the distortion level $D = Ed(X^n, \hat{X}^n)$ achieved by the chosen code. The proof of the converse is omitted here.

Bibliography

1. R, G. Gallager, *Information Theory and Reliable Communication*, John Wiley & Sons, Inc., New York, NY, 1968.
2. Thomas M. Cover and Joy A. Thomas, *Elements of Information Theory*, Wiley-Interscience, 2006.
3. T. Berger, *Rate Distortion Theory*, Prentice-Hall, 1971.
4. J. P. Crutchfield and D. P. Feldman, "Regularities unseen, randomness observed: Levels of entropy convergence," *Chaos: An Interdisciplinary Journal of Nonlinear Science*, 2003, pp. 25-54.
5. J. D. Gibson, "Mutual Information, the Linear Prediction Model, and CELP Voice Codecs," *Information*, Vol. 10, No. 5, 2019.
6. J. P. Crutchfield and D. P. Feldman, "Synchronizing to the environment: Information-theoretic constraints on agent learning," *Advances in Complex Systems*, Vol. 4, 2001, pp. 251-264.
7. C. E. Shannon, "A mathematical theory of communication," *Bell Sys. Tech. Journal*, Vol. 27, 1948, pp. 379-423.
8. J. Gibson, "Speech Compression," *Information*, Vol. 32, No. 1, 2016.
9. R. M. Gray, "A new class of lower bounds to information rates of stationary sources via conditional rate-distortion functions," *IEEE Trans. on Information Theory*, Vol. IT-19, No. 4, July, 1973, pp. 480-489.
10. L. D. Davisson, "Rate-distortion theory and application," *Proceedings of the IEEE*, Vol. 60, No. 7, July 1972, pp. 800-808.
11. R. M. Gray, "Information rates of autoregressive processes," *IEEE Trans. on Information Theory*, Vol. 16, No. 4, July 1970. pp. 412-421.
12. C. E. Shannon, "Communication in the presence of noise," *Proceedings of the IRE*, Vol. 37, No.1, 1949, pp. 10-21.
13. E. L. Lehmann, *Theory of Point Estimation*, John Wiley and Sons, Inc., 1983.
14. E. L. Lehmann, *Testing Statistical Hypotheses*, Second edition, John Wiley and Sons, Inc., 1986.
15. S. Still, "Information-theoretic approach to interactive learning," *EPL*, Vol. 85, Jan. 2009.
16. S. Still and D. Precup, "An information-theoretic approach to curiosity-driven reinforcement learning," *Theory Biosci.*, Vol. 131, Sept. 2012, pp. 139-148.

J. D. Gibson, *Information Theoretic Principles for Agent Learning*, Synthesis Lectures on Engineering, Science, and Technology, https://doi.org/10.1007/978-3-031-65388-9

17. J. P. Crutchfield and C. R. Shalizi, "Thermodynamic depth of causal states: Objective complexity via minimal representations," *Physical Review E*, Vol. 59, No. 1, Jan. 1999, pp. 275-283.
18. S. Still, J. P. Crutchfield, and C. J. Ellison, "Optimal causal inference: Estimating stored information and approximating causal architecture," *Chaos*, Vol. 20, Sept. 2010.
19. N. Tishby and D. Polani, "Information Theory of Decisions and Actions", in Cutsuridis, V., Hussain, A., Taylor, J. (eds) *Perception-Action Cycle*, Springer Series in Cognitive and Neural Systems. Springer: New York, NY, 2011.
20. O. Shamir, S. Sabato, and N. Tishby, "Learning and generalization with the information bottleneck," *Theoretical Computer Science*, Vol. 411, 2010, pp. 2696-2711.
21. N. Tishby, F. C. Pereira, and W. Bialek, "The information bottleneck method," in *Proc. 37th Annual Allerton Conf. on Communication, Control and Computing*, 1999, pp. 368-377.
22. N. Tishby and N. Zaslavsky, "Deep learning and the information bottleneck principle," 2015 IEEE Information Theory Workshop (ITW), 2015, pp. 1-5.
23. B. Rimoldi, "Successive refinement of information: Characterization of the achievable rates," *IEEE Trans. Information Theory*, Vol. 40, No. 1, Jan. 1994, pp. 253-259.
24. W. H. R. Equitz and T. M. Cover, "Successive refinement of information," *IEEE Trans. Information Theory*, Vol. 37, No. 2, March 1991, pp. 269-275.
25. R. E. Blahut, *Principles and Practice of Information Theory*, Addison-Wesley Longman Publishing Co., Inc., 1987.
26. T. Berger and J. D. Gibson, "Lossy Source Coding," *IEEE Trans. on Information Theory*, Vol. 44, No. 6, Oct. 1998, pp. 2693-2723.
27. R. McEliece, *The Theory of Information and Coding*, Cambridge University Press, 2002.
28. A. Kolmogorov, "On the Shannon theory of information transmission in the case of continuous signals," *IEEE Trans. on Information Theory*, Vol. 2, No. 4, 1956, pp. 102-108.
29. S. Kullback, *Statistics and Information Theory*, J. Wiley and Sons, New York, 1959.
30. M. S. Pinsker, *Information and Information Stability of Random Variables and Processes*, San Francisco, CA, USA, Holden-Day, 1964.
31. A. El Gamal and Y.-H. Kim, *Network information theory*, Cambridge University Press, 2011.
32. J. J. Shynk, *Probability, random variables, and random processes: theory and signal processing applications*, Hoboken,NJ, USA, John Wiley Sons, 2012.
33. D. G. Messerschmitt, "Accumulation of Distortion in Tandem Communication links," *IEEE Trans. on Information Theory*, Vol. (T-25, No. 6, Nov. 1979, pp. 692-698.
34. J. Gibson and H. Oh, "Analysis of Cascaded Signal Processing Operations Using Entropy Rate Power," Asilomar Conference on Signals,Systems and Computers, 2018.
35. J. D. Gibson, "Log Ratio of Entropy Powers," UCSD Information Theory and Application Workshop, 2018.
36. J. D. Gibson "Entropy Power, Autoregressive Models, and Mutual Information," *Entropy*, 2018.
37. P. Elias, "A note on autocorrelation and entropy," *Proceedings of the Institute of Radio Engineers*, Vol. 39, No. 7, 1951, p. 839.
38. W. Ebeling, "Prediction and entropy of nonlinear dynamical systems and symbolic sequences with LRO," *Physica D: Nonlinear Phenomena*, Vol. 109, No. 1-2, 1997, pp. 42-52.
39. J. D. Gibson, "Mutual Information Gain and Linear/Nonlinear Redundancy in Agent Learning, Sequence Analysis, and Modeling," *Entropy*, May 2020.
40. J. E. Hudson, "Signal Processing Using Mutual Information," *IEEE Signal Processing Magazine*, Vol. 23, No. 6, Nov. 2006, pp. 50-54.
41. L. R. Rabiner and R. W. Schafer, *Digital processing of speech signals*, Pearson, 2011.
42. J.-H. Chen and J. Thyssen, "Analysis-by-Synthesis Speech Coding," *Springer Handbook of Speech Processing*, Springer, 2008.

43. G. A. Darbellay and I. Vajda, "Estimation of the information by an adaptive partitioning of the observation space," *IEEE Transactions on Information Theory*, vol. 45, No. 4, May 1999, pp. 1315-1321.
44. C. E. Shannon, "Coding theorems for a discrete source with a fidelity criterion," *IRE Nat. Conv. Rec*, Vol. 4, 1959, pp. 142-163.
45. L. Rabiner and B.-H. Juang, *Fundamentals of Speech Recognition*, Prentice-Hall, Englewood Cliffs, N. J., 1993.
46. D. O'Shaughnessy, *Speech Communications: Human and Machine*, IEEE Press, 2000.
47. B. W. Silverman, *Density estimation for statistics and data analysis*, Routledge, 2018.

Bibliography